FastTrack © Project Management

FastTrack © Project Management

Lee Lister is a Program and Bid Manager with more than 25 year's management and consultancy experience and 20 year's program and project management experience in projects for many household names. She also has 15 years bid management experience ranging from bids for medium companies to large international and country infrastructure bids.

On the internet she is known as **"The Bid Manager"** or **"The Biz Guru"**.

Whilst working in the Far East she became a recognized expert on preparing and evaluating large World Bank Proposals (infrastructure and business process projects within developing countries). She also consulted on setting the World Bank Bid Evaluation Criteria. This expertise was acknowledged by an invitation to be the principle speaker at an International Business Development Conference in Washington, USA.

She has also consulted at very senior and level and with government officials in several countries in Asia, Europe, America and Australia. She has set up project and program management environments for many large blue chip companies all over the world. She has personally managed in excess of 50 projects and 20 programs of work. Her experience encompasses, bid management, bid evaluation, negotiation, bid training, consultancy, project and program management.

She is a prolific published writer of books, ebooks and articles and can easily be found on major search engines.

FastTrack © Project Management

First published in Great Britain in 2008. Previous incarnations were used as training materials from 1998.

© **Copyright Lee Lister 2008**
All rights reserved.

Photo Copyright © Pressmaster

No part of this publication may be reproduced, stored in a retrieval system, or transmitted in any form or by any means, without the prior permission in writing of the publisher, nor be otherwise circulated in any form of binding or cover other than that in which it is published and without a similar condition including this condition being imposed on the subsequent purchaser. This book may not be used as a training course in any format.

ISBN: 978-1-907551-01-7

Other books available include:
FastTrack© Project Management
FastTrack© Bid Management
FastTrack© Finding The Winning Solution
FastTrack© To Job Success
Consultant's Tool Box

FastTrack© Project Management

Understand how to manage a project effectively, efficiently and profitably. Understand project controls, documentation, planning, management and communication with our detailed methods and processes.

www.ProjectNiche.com

This book is dedicated to my daughter Kerry Lister for whom I have always strived to be my best.

FastTrack © Project Management

CONTENTS

LEGAL NOTICE	**10**
INTRODUCTION	**11**
Some Useful Definitions	11
What does a Project Manager do?	13
Project Strategy	15
Risk Levels	*15*
Why do Things go Wrong?	16
Project Initiation	*16*
Project Duration:	*17*
Project Implementation:	*19*
The Essence of a Good Project	20
THE PROJECT TEAM	**21**
Project Control Office	22
Quality Review Team	23
Specialist Consultants	23
Business Analysts and Systems Analysts	23
Programmers	25
Trainers	26
Testers	26
Implementation Team	27
Maintenance Team	28
Motivating the Project Team	29
What can you do to motivate staff?	*30*
Charge out rates for staff	31
Choosing your team member	33
PROJECT STAGES	**35**
Why Divide a Project into Stages?	35
The Main Project Stages	36
Project Reports	37
Project Inception	37
Feasibility	39
Project Planning	41
Quality Assurance	44
Analysis	45
Design	47
Development	48
Testing	50
Implementation	52

FastTrack © Project Management

Data Conversion and Take on	53
System Support	54
User Support and Training	55
User Acceptance	56

PROJECT PLANNING — 64

Why Project Plan	67
Problems of Project Planning	67
Planning Tools	68
Steps to a Project Plan	68
Transition Planning	70
Budgets	70
Steps to Planning a Manpower Budget	*72*
An Example of a Budget	*73*
Steps to planning a Resource Budget	74

PLANNING DOCUMENTS — 75

Project Plan	75
Staffing Plan	76
Work Allocation	76
Budget	77
Quality Plan	78
Test Plan	79

CONTROLS WITHIN A PROJECT — 81

What Types of Changes are There?	81
Controls in Action	82
Why Control Them?	82
Effects of Lack of Controls	83
Summary	84
Change Control	85
Overview	*85*
When Change Control Should be Used	*85*
Aims of the Change Control Procedure	*85*
Actions	*86*
Quality Control	88
What is Quality?	*88*
Quality Control	*89*
Quality Strategy	89
Quality Plan	*90*
Product Descriptions	*90*
Quality File	*91*
Quality Reviews	*91*
Risk Control	93

Areas of Risk	93
Risk Assessment	93
How to Minimise Risk	94
Requirements to Ensure Minimum Risk	94
To Minimise Risk.....	95
Security	95
System Environment	95
Hardware	96
Software	97
Staff	98

IMPLEMENTATION CONTROLS   99

The Implementation Procedure	100
Change Management	101
Change Management Problems	101
Post Implementation	102
Version Control	103
Overview	103
Release Numbering	103
Deliverables Schedule	104
Version Isolation	104
Release Control	105
Testing Control Procedure	106
Overview	106
Unit Testing	107
System Testing	108
User Acceptance Testing	109
Error Control	109
Implementation Control Reports	111
Change Control (CRF)	111
Error Report	112
Version Control Report	113
Release Control Report	113
User Acceptance	115
Overview	115
Why is it so Important?	115
The Procedure	115
User Acceptance Sign Off Report	116
Requirement Control Reports	117
Project Initiation Document (PID)	117
User Requirements Specification	119
System Specification	121
Functional Specification	122
Program Specification	122

Requirement Control Reports ... 123
   Project Initiation Document (PID) ... *123*
User Requirements Specification ... 125
System Specification ... 127
Functional Specification ... 128
Program Specification ... 129

COMMUNICATION ... 130

With your Clients ... 130
   Consultancy ... *130*
   Client Relationship ... *130*
   Steering Committee ... *130*
   Project Progress Reports ... *131*
With your Project Staff ... 131
   Project Reports ... *131*
   Project Meetings ... *132*
   Time and Work allocation ... *133*
   Staff Reviews and Meetings ... *133*
With the Users ... 133
   User Meetings ... *134*
   User Questionnaires ... *134*
Progress Reviews ... 135
Project Inception and Set up ... 137
   Project Set up ... *137*
   Objective setting ... *138*
   Project Charter ... *139*
Communication Reports ... 139
   Terms of Reference ... *139*
   Statement of Scope and Objectives ... *140*
   Project Charter ... *141*
   Project Progress Report ... *144*
   User Questionnaires ... *145*
   Test Acceptance Report ... *145*
   User Acceptance Test Report ... *146*

THE IMPORTANCE OF DOCUMENTATION ... 147

Analysis and Design - Standard Procedures ... 148
System Support and Maintenance ... 150
Development and System Support ... 150
User Support ... 151
Technology Transfer ... 151
Training ... 153

PROJECT DOCUMENTATION REPORTS ... 154

Analysis Reports	154
Document Flow Diagram	*154*
Data Flow Diagram	*155*
Entity Relationship Diagram	*156*
Entity Life History	*157*
System Reports	158
Program Documentation	*158*
Data Dictionary	*158*
Management Reports	159
Management Overview	*159*
User Documentation	160
Help Screens	*160*
User Handbook	*160*
User Guide	*161*
Procedures Guide	*161*
Maintenance and Support	162
Procedure Documentation	*162*
System Documentation	*162*
User Support Documentation	163
Operations Guide	*163*
Installation and Maintenance Manual	*164*
JARGON BUSTER	**166**
INDEX	**186**

TABLE 1 HERBERG'S THEORY 29
TABLE 2 STAFF MOTIVATING 30
TABLE 3 CHARGE OUT RATES 33
TABLE 4 SUMMARY OF DOCUMENTS PRODUCED 58
TABLE 5 GENERIC PROJECT PLAN 60
TABLE 6 PROJECT PLANNING MODEL 64
TABLE 7 OBJECTIVES 138

Legal Notice

We do not believe in get rich quick schemes. We do believe that business is equal parts of inspiration, hard work and luck. Every effort has been made to accurately represent our product and it's potential.

Please remember that each individual's success depends on his or her background, dedication, desire, and motivation. As with any business endeavor, there is an inherent risk of loss of capital. **There is no guarantee that you will earn any money**.

This book will provide you with a number of suggestions you can use to better guarantee your chances for success. **We do not and cannot guarantee any level of profits.**

This product is written with the warning that any and every business venture contains risks, and any number of alternatives. We do not suggest that any one way is the right way or that our suggestions are the only way. On the contrary, we advise that before investing any money in a business venture you seek counseling and help from a qualified accountant and/or attorney or lawyer.

> **You read and use this product on the strict understanding that you alone are responsible for the success or failure of your business decisions relating to any information presented by our company Biz Guru Ltd.**

FastTrack © Project Management

Introduction

A Project is a group of defined and linked tasks that allow, using a formal methodology and a series of controls, a pre-determined business requirement to be fulfilled, by a series of formal deliverables within the constraints of time, resources, budget and quality.

Some Useful Definitions

All Projects have at least one of the following four aims:
- To save manpower or other resource costs
- To improve customer service
- To improve the collection, collation control and/or dispersion of information
- To allow an organization to make innovative or unique changes or strides within their environment

Tasks: One or a group of actions that must be undertaken in order to achieve or produce something. Each task has a well defined entity, and is often logically linked by action or time.

Methodology: A structured and formal method of planning, managing, controlling, documenting and reporting upon a project. The methodology is used as a common language amongst the project team. It contains modules that define how to approach each stage of the project.

Controls: The methodology explains how to undertake a project. Controls are used to ensure that the project is proceeding as required.

Business Requirement: The definition of the problem faced by a business and the justification for requiring the system, an outline of the benefits which will accrue, the savings it will bring.

Resources: People, equipment, office space, computer time, expert time, etc. which are to be availed the project.

Time: That of which there is never enough in any project.

A **Baseline** for each of the following factors is established, from which variances and their ramifications are monitored and measured. These baselines form the basis of the reporting structure both internally and externally to the project team: Quality, risk, benefits, change, resources, budgets, project progress, deliverables and configuration versions.

Business Case: The justification for undertaking a project, defining the benefits that the system is expected to deliver and the savings it will accrue, measured against the cost and the risks of implementing and running the system.

Deliverables: A tangible, important part of the resulting system, the implementation of which marks a milestone within the project plan. Major Deliverables are often part of the contract between the project team and the users. All deliverables should feature within the project plan. An example is a major report, a software module or a computer sub-system.

Quality: British Standard 4778 quotes "...the totality of features and characteristics of a product or service which bears on its ability to satisfy a given need." A quality system, an often forgotten, but very important deliverable of a project is:
- Error Free
- Well structured and defined
- Complete and Comprehensive
- Flexible
- Efficient
- Reliable
- Easy to maintain
- Easy to enhance
- Well documented

Budget: The money allocated to the project. This has two constituents -- Capital and Income, sometimes called Non Recurring and Recurring Costs. Capital costs are one-off costs on significant and tangible items, e.g. a computer. Income costs are often for intangible items and expected to be recurring, often beyond the life of the project, e.g. hardware maintenance.

What does a Project Manager do?

- Defines and reviews the Business Justification and Requirements, by regular reviews and controls, to ensure that the client receives the system that he wants and needs.
- Instigates and plans the project by establishing its format, directions and base lines that allow for variance measurements and change control.
- Controls the constituents, progress and direction of the project by achieving goals, reaching targets and delivering on time.
- Manages technology, people and change in order to achieve goals, reach targets and deliver on time.
- Manages the Project Staff by creating an environment conclusive to the delivery of the new application in the most cost effective and effective manner.
- Manages the Client Relationship by using an adequate complete and formal reporting format that compliments a respected and productive informal relationship.
- Drives the project by leading, pushing and motivating whilst ingesting stress and pressures until the project reaches the required goal.

What Qualities does a good Project Manager need?

- Have a recognized and respected senior position.
- A strong, determined but still approachable personality
- Be a persuasive communicator
- Be sensitive to the problems of change
- Quick decision maker
- Technically competent
- Effective planner
- A good controller
- Good team manager
- Visionary, but still able to attend to minor details and multiple ramifications.
- Able to work in stressful environments

Why Does it Look so Easy? Project Management is a professional skill that is often not noticed. A good project manager will be organized and have his working tools, such as his project plan and standard documents set-up to assist with control, communication, reviewing and management of the project. When a project experiences problems (and most do) these same tools, combined with his skill and experience will allow for forewarning of problems ahead and a speedy resolution to allow the project to come back onto course.

The skills that a good project manager will have are:
- His skill and experience.
- His personae or professional image, emits confidence and ability
- Determination and stubbornness tendered with patience and consistency
- Non- obtrusive project management skills and management style
- Formal Training, both of the Project Manager and the project team
- Good, firm and structured controls are established within the project plan
- People know what their aims and responsibilities are
- He picks the right team

The tools that he would use are:
- Project planning tools and methodology
- Communication infrastructure
- Team management
- A set of effective project controls
- Set of standard documents
- Office software including diary

Project Strategy

Whilst planning your project and gathering your business case and business objectives together it is important to plan your project strategy. The strategy will include factors such as the emphasis you place on various objectives and defining the amount risk that your client wishes to face.

The "Two out of Three Rule"

Pick two out of the following: **QUALITY   TIME   PRICE**

A quick examination of the above three criteria will indicate any choosing any two will adversely affect the third option. Different projects will place different emphasise on each of the above. How a project is managed depends upon this emphasis. Do you prioritise on time and risk quality? Do you undertake a fixed price and time project and risk a compromise on quality? It takes a very special skill and an ability to foresee all unknown factors and business influences that the project may face in its life time.

Risk Levels

Early on in your discussions with your client you will become aware of the level of risk that the business and the client are able to face, with respect to implementing the system. For mission critical systems (e.g. air traffic control, dealing desks), or risk adverse companies (e.g. retail banks) the level of risk would be very low. A company struggling to get a new product to market ahead of its rivals may accept a high level of risk (backed by contingencies and upgrade plans) in order to reap the benefits of being "the first". The level of acceptable risk will affect, factors such as the:
- degree of controls required.
- volume and frequency of progress reviews and reporting.
- complexity of the project plan.
- cost of the project.
- number of resources, including team members, required.
- technical solution (e.g. package software vs. new build).

Why do Things go Wrong?

With all the project tools and skills available what can go wrong and why?

Project Initiation

Pre-conceived ideas as to the solution - one of the biggest sins a consultant or Project Manager can have it to define a Business Requirement of a solution around a technical platform, software package or operating system, instead of the other way around. It is important that a consultancy has a number of different possibilities at its fingertips. This is a particular problem for hardware manufacturers who also implement full systems. Open systems are making this a less common problem.

Poor analysis or visualization of client's needs - the client's needs have to be fully analyzed, documented and agreed by all. This manages client expectations and ensures that the project has a firm base from which to proceed.

Unreasonable expectations from client and users of the system. At project initiation it is very important to establish the reasonableness of the client's request, and also to ensue that what they are requesting is indeed what you are going to give them. This misunderstanding at the crucial instigation of the project can mean that although resulting system is technically a success, it is never used because it isn't what was wanted. In other cases the system is used but myriads of manual systems are used to by pass the idiosyncrasies of the new system. It is extremely important to ascertain and document the client's requirements and expectations at an early stage. All of the project reports dealing with these matters have User's Authorizations attached to them, for just this reason.

Lack of support and/or interest - often a project is viewed by users as the "pet" idea of the bosses. The Project Manager has to instil enthusiasm and interest in order to ensure the success of the project.

Unreasonable expectations from client and users of the budget and resources - often clients have very little idea of how much their ideas and requirements will cost, neglect to budget for contingencies, and most commonly, forget that major changes take resources, time and budget. It is important that at the beginning of the project and at each Change Control that accurate estimate of budget, time and resource requirements are given and formally agreed. The Project Review report should include up to date costs and forecasts.

Risks, costs and benefits incorrectly measured -they can soon become a problem and threaten even the best managed projects, either because or a drain on resources or because the client's expectations have not be properly measured and controlled. All possibilities should be explored during the Feasibility stages.

Project Duration:

Poor Project Management- Reading through this book will highlight just how complex and wide the skills of project management are. Even if someone knows all the techniques, the rapidly changing fluidity of one or more concurrent projects means that there is no substitution for experience. The misuse to the controls and planning techniques can hamper or confuse the smooth running of the project. An improperly set-up project has very little chance of success as the project's rapid evolution quickly compounds the problems, such that no-one is really sure of what they are doing or what has been done.

Lack of controls - Many projects fail, go over budget, or run out of time, simply because resources, change, quality and risk have not been properly measured, monitored and controlled. A common mistake is not to establish a base line or control group, such that the ever changing expectations and results quickly get out of hand.

Poor communication- internal or external to the project means that no-one is really sure what is happening or meant to happen.

Quality not measured or controlled - a quality controlled project ensures that it will work, be supportable and efficient.

Poorly developed systems planning mechanisms - the appropriate methodology must be chosen for the type and size of project. Too much reporting and project management means that nothing much gets done in a sea of paper. Too little means that problems happen and controls slip without the Project Manager being aware or able to compensate or rectify.

Inappropriate project management methods - the method used by the Project Manager may not be correct for the system, project team or culture of the company, such that the monitoring, controlling and reporting mechanisms become ineffective or inefficient.

Inappropriate or ineffective organizational structures - unless the correct organizational place for the Project Manager, project team, project and system can be found, obtaining agreements, sanctions and support can become a nightmare.

Loss of project momentum - in the middle of the project, when the excitement the start has abated, and before an end can be seen, the project frequently loses impetuous and controls and enthusiasm starts to wane. It is at this time that the Project Manager needs to be extra vigilant and enthusiastic.

Lack of training and formal skills - More than any other profession, Information Technology (IT) has more than it's fair share of "experts" who have limited or narrow knowledge of a very complex subject, that is continually liable to complex changes and enhancements. Rapid evolution of the IT profession has meant that there is a lack of formal qualifications and structured training such as you will find in the other major professions such as law and architecture. The strange jargon, that is necessary part of working within the profession, confers expert status upon someone who can converse using a few difficult to understand words.

Inappropriate mix of skills in project team - not only does the project team have to have the full range of skills, but each member should compliment and enhance the others. Sorting out inter team rivalries and why something has not been done is not a constructive way for a Project Manager to spend their time.

Low confidence on in-house computer group - they often bear the brunt of the user's criticisms and complaints about the existing systems. It is important to train and support this group, gaining its confidence and backing.

Low confidence in project team - a team that the clients or users don't believe will achieve the project's aims.

Project Implementation:

Resistance to change - you will always hear... We always do it this way" from someone. The art is to overcome this resistance, which is one of the biggest killers of a successful system implementation.

Suspicion or conflict - a good Project Manager must use all his skills and personality to overcome these problems.

Inadequate analysis or testing - a sure fire way to obtain an error ridden system that does not meet the user's requirements. Although it is sometimes tempting to cut short the more boring parts of a project, consistent and complete analysis and testing, saves much work and cost at implementation. A badly written or bugged piece of software is very bad for the moral of the users and confidence of the clients in the consultancy.

System collapse after implementation - through poor technology transfer, poor documentation, poor or inappropriate training, is a common source of failure. The system has to support by the client and the users after implementation. People often find a way of overcoming the failings of a badly designed system, especially when it fails to adapt to required changes. It thus loses user support and approval, who then devise alternative and unofficial systems to cope with the new requirements.

The installed system then becomes by passed, despised and falls into disuse.

The Essence of a Good Project

Throughout this book it is intended to detail how you can set up you own project tools and gain project management skills that will allow you to successfully manage a project. This book will build upon the following four main core skill requirements:

Plan

Control

Communicate

Document

The Project Team

The *Project Manager* leads, manages and controls the entire project team. He also designs, writes and controls the project plan and manages the client relationship. In smaller projects he undertakes the Technical Co-ordinator and Application Software Managers role, but never the Quality Manager's role.

The *Quality Manager* is not a member of the project team as he reports to the Steering Committee and not the Project Manager. This role is undertaken by the client's own quality review department or by an independent external consultancy. He is responsible for the *independent* quality reviews and controls within the project. Independence and an attention to detail are the qualities most needed for this position.

The *Technical Co-ordinator*, works only on the hardware side of the project if there is an Application Software Manager in the team, otherwise he undertakes both software and hardware coordinating. He is responsible for making sure that the hardware and/or software components of the system are installed and implemented in the right place at the right time. This would encompass liaising with site preparation, testing, programmers, users and clients. This person needs to be organized, determined and tactful.

An *Application System Manager* is included on the team when there is a significant amount of software to be written and/or installed. The significance of the software may be in its volume or its importance. He is responsible for the entire production and implementation of the software. He works closely with the rest of the Project Management team, the analysts and programmers and testers. Excellent analysis skills and programming knowledge is required. He does not need to be a programmer but certainly needs substantial analysis and consultancy skills.

A *Project Leader,* of which there may be one or many, is in charge of a small group of staff, a specific function (e.g. systems analysis) or a specific task or tranche (e.g. testing or implementation of a software module). He works directly to the appropriate senior manager and has line management responsibilities. Sometimes he also has budgetary and control responsibilities and reporting requirements. He needs to be very skilled within the function that he leads and also requires basic project management and good leadership skills. The post of Project Leader is often that of a learning ground for potential Project Managers, or alternatively the position may go to a very experienced analyst or programmer. The role of the Project Leader is to assume some of the less complex project management duties.

A good Project Manager will delegate to their leaders in such a way that the leader has their skills developed whilst the Project Manager's time is dedicated to the more pressing and technical problems. The Project Leaders are sometimes included within the Project Management team and sometimes within each separate function.

Project Control Office

The *Project Control Office* (PCO), working for, and very closely to the Project Management Team, have the following five functions:
- **Communication Centre** - All information with respect to the project, the project plan, its reports and its staffing is generally fed through the PCO, where the appropriate people are notified and action taken.
- **Reporting** - All the reports, external and internal to the project team are produced within the PCO. Depending upon the size of the project, the preference of the Project Management Team and the skills of the staff within the office, the reports will either be written and produced by the PCO or merely collated, distributed and stored.
- **Controlling** - The accounts and controlling documentation are updated and stored.

- **Secretarial and clerical** - The secretarial staff for the Project Management Team and the clerical staff for the project generally reside within this office.
- **Librarian** - The multitudes of reports, supporting documentation, software releases, manuals, training materials etc. appertaining to the project and the ongoing support of the resulting system are usually managed and controlled from within the PCO. The physical storage often happens elsewhere.

Quality Review Team

This team, headed by the *Quality Manager* are responsible for ensuring that the quality of each individual constituent of the project, project reporting and resulting system meet the pre-determined standards. This is undertaken by regular auditing, measurement and monitoring of inputs and outputs, the system, control methods and reporting functions.

The *Quality Review Team* will consist of members representing the interests of the users, the internal project management team, and an independent external body, such as a consultancy or the company's own quality review function.

Specialist Consultants

Many projects require specialist consultants for a short period of time, either to advice upon specialised problems or areas of implementation or analysis. Such specialists may be financial, taxation, legal or accountancy or more specific to a particular piece hardware or software.

Business Analysts and Systems Analysts

The difference between a Business and a Systems Analyst causes much confusion. In some small projects business and systems analysis is undertaken by the same person, but more often the Project Manager undertakes the business analysis and the Programmer undertakes the systems analysis.

The *Business Analyst* interfaces and interacts between the user and the systems analyst and programmer. His work is concentrated at the beginning of the project and before the programming begins. He will analyze and document the existing system and system requirements, by the use of User Questionnaires, Workshops, interviews, Walkthroughs, Flow Charts, Document Flow Diagrams, Data Flow Diagrams (DFD), Workflow Diagrams, Entity Relationship Diagrams (ERD), and User Requirement Specification (URS). Although the documents and diagrams will be of a technical nature, they should be written and explained with the user in mind. He often produces the User Documentation, although this is sometimes undertaken by the Trainers.

The *Business Analyst* needs to have a considerable amount of business, consultancy and analysis experience. The ability to meet and talk to people in a natural and detailed way whilst still gathering and collating the required information is a must. A detailed knowledge of IT and its potential is also required. The ramifications on the proposed hardware and software configuration of the Business Requirements need to be understood. The production of the various documents and diagrams requires formal training, and it is the ability to produce, explain to non technical people, and use these methodologies that separates the skilled Business Analyst from the many people that purport this knowledge. Even more than that of a Project Manager the Business Analyst is the IT job that has the most number of untrained and unsuitable incumbents.

The *Systems Analyst* interfaces between the Business Analyst and the Programmers. He will take the documentation produced by the business analysis and document it into a format suitable for the programmer to produce a system. He will produce the following documentation: System Specification, Functional Specification, Entity Life History (ELH), Physical Data Design, Logical Data Design, Logical Entity Relationship Diagram (LRD) and System Documentation. This documentation is very technical and written especially for internal use, particularly for the programmers.

The *Systems Analyst* needs to have considerable programming skills, and although not necessarily a programmer, he must have extensive technical skills and knowledge, particularly with respect to data and software structures. The documentation needed is very technical and detailed such that formal training is needed to accomplish this. His work is concentrated in the middle of the project. Often the Systems Analyst will assist is setting up the test specification.

The *Systems Designer* is a highly skilled and experienced programmer and systems analyst. It is his job to design the databases and file structures to afford the optimum performance of the resulting system. He concentrates more on the very specific and detailed design rather than analysis work. You will find a Systems Designer on a project team when a very large and complex database or information system is required. He will often take the work from the systems analyst and produce more detailed designs, taking into account the performance of the specific hardware, operating systems and software packages.

Programmers

The *Programmer* cuts code, writes the software and designs and writes the software modules and interfaces as required. He is also responsible for producing the Data Dictionary, Program Specification, System Documentation and Version Control Documentation.

Programmers have a reputation of sometimes being a little eccentric. Certainly in the past, Programmers were often shut away in a room, keeping very strange hours doing nothing but cutting code that was so complex that it held little interest to outsiders. The more wide spread use of CASE Tools, RDBMS, Client-Server architecture and Dos based programming, has removed much of the monotonous work, in favour of more technical assignments. A Programmer needs to be detailed and structured in his work and prepared to fully document and test the programming. His skills need to be continually updated in line with the latest technology and software version releases. Emphasis is now upon building programs from modules and standard interfaces.

Programmers tend to specialise on certain hardware platforms or major software packages. Some Programmers also concentrate on software maintenance as opposed to writing new programs.

An *Analyst/Programmer* also undertakes the Systems Analyst role, whilst a Programmer/Analyst concentrates mainly on programming but assists the System Analysts. In the salary stakes the former is paid more than the later and both more than a Programmer.

Trainers

Trainers, led by the Training Manager, design, write and undertake the training of the users, to ensure that they can use and maintain the new system after its implementation. Training usually starts as implementation begins.

*Trainer*s need to be patient, organized and authoritative. The ability to produce interesting and informative training material from a Desk Top Program (DTP) environment is also necessary. Trainers often work anti-social hours and are subject to some resistance from disinterested or apprehensive users. A sense of humour and persistence is also required, as is the ability to explain technical matters in a non technical manner.

Testers

There are three types of testing:
- **Unit Testing,** usually conducted by the programmer responsible for production of the unit under test; generally the person who has programmed the unit. The purpose of unit testing is to qualify individual components of a system. The tester ensures that the unit meets the program specification.
- **System Testing**, based on the Test Plan which is prepared in the Design phase of the project life cycle. The purpose of system testing is to qualify the system as a whole, by testing the menu structures, user and program interfaces, etc... The tester may be anyone who has not programmed the part of the system under test.

- **User Acceptance Testing**, based on the Test Plan which is prepared in the Design phase of the project life cycle. The purpose of user acceptance testing is to qualify the system as a whole, to the satisfaction of the Project Sponsor, ensuring that the Project Objectives have been met and the User Requirements satisfied. This testing will be undertaken by the users, but the testing team will usually advise and guide as to the mechanics of the tests.

Testers work alongside the analysts and programmers, although to maintain their independence, it is important that a detailed knowledge of the program is not obtained. This is to ensure that the results of the tests are what they see and not what they expect to see from their knowledge of the program. *Testers* will also work alongside the Quality Review Team, comparing results.

The *Testers* will undertake system testing, supervise the programmers testing and offer consultation on user testing. They must resist the ever present pressures to also undertake user testing.

Testers need to be patient and meticulous, working in a logical and constraining environment, whilst maintaining their independence during a period when everyone is keen to move onto implementation.

Implementation Team

The *Implementation team,* led by an Implementation Manager, are responsible for the installation of hardware, software and comms within the prescribed site. They liaise with the Technical Coordinator, Application System Manager and Testers to ensure that the relevant equipment and software are ready.

The *Configuration team* will install and configure the:
- hardware, operating systems and hardware support software (e.g. Unix, remote access, hardware optimization packages)
- network operating system
- peripherals
- package software

Site Preparation is often included within this team. This group ensure that the physical locations are suitable for the hardware, software and comms installation. Cablers, electricians and carpenters need to be available. Cool air, electrics, communications, fire prevention and safety codes are the requirements of site preparation. Some hardware, e.g. dealer desks, needs to be installed on a hollow floor, to ensure that all the cabling is optimally run and stored.

Maintenance Team

There are four types of maintenance that need to be instigated, at the end of the project. Maintenance is usually controlled by contract, giving details of minimum response times and geographic or system coverage. It has penalty clauses for non performance. This contract is usually called a Support Contract or Service Level Agreement (SLA). It is standard procedure within the Computer Industry to charge a fixed % of the hardware, software or comms as the maintenance charge.

Hardware maintenance is to support and make any upgrades and changes necessary to the hardware, its peripherals and operating systems. It usually takes the form of a support help desk, backed up by on-site or off-site engineers, who have access to a spares store. Many hardware systems have built in software packages that monitor and report on the configuration and its operating systems.

Comms maintenance is similar to Hardware maintenance but the network configuration and communication software is often included within the maintenance.

Software maintenance usually takes the form of a hot line telephone support and at least one but usually two or three escalation levels backing it. Support takes the form or user assistance, bug fixes and version upgrades. Many help desks now have modem access to the system to facilitate faster service.

Documentation maintenance is different in that it rarely has hot line support. The maintenance takes the form of regular documentation and manual updates. The system documentation also includes the help screens and program annotations, which are updated under the software contract. Documentation maintenance is often included within the software maintenance contract.

To work in any of these fields, the person must be technically proficient and able to work the unusual hours demanded. As the people are usually on-site a presentable personality and persona is also required.

Help Desk work requires the ability to make quick decisions form the little information presented. A thorough knowledge of the supported product and the ability to work under very stressful conditions is vital. Junior members of staff are often started on the Help Desk to ensure familiarization of computing work.

Motivating the Project Team

Having decided upon the make up of your team, your staff have to be motivated in order for them to give their best and work happily together to the best advantage of users, clients, project and staff. The most common theory of motivation used is...

Table 1 Herberg's Theory

Highest	Motivation Factors	Achievement
		Recognition
		Work itself
		Responsibility
		Advancement
	Hygiene Factors	Company policy and administration
		Supervision - technical
		Supervision - Interpersonal relationships
Most Basic		Salary
		Working conditions

This theory reflects that once the "Hygiene" factors of a happy, safe working environment have been met, then the worker will look to job related motivation factors to meet his needs.

The "Hygiene" factors are not motivators as such, but make people dissatisfied when they are not properly present. Clearly the working environment must be comfortable and appropriate before workers will experience any motivation from the job content.

Table 2 Staff Motivating

Factor	Motivating	De-motivating
The Job	Interesting Stretching Area of Responsibility	Impossible Too easy or too routine Poor remuneration
Achievement	Identified and achievable goals	Little support or guidance Lack of recognition Negative criticism
Advancement	Possibility of development Promotion	Lack of responsibility Restrictions on authority
Work Environment	Safe Happy Pleasant	Threatening Lack of communication Confusing team structure

What can you do to motivate staff?
- Observe your team member's behaviour and interaction
- Listen to what they tell you
- Communicate theirs and the project's aims and objectives

Act to remove de-motivating factors by giving more responsibility for:
- People
- Finance
- Planning
- Actions
- Communication
- Training

Ensure that any tasks given to team members are:
- Complete and Entire
- Varied
- Given with the authority to make decisions
- Increasingly more challenging
- Well defined
- Achievable within the abilities of the staff member

Conclusion

☞ Know your staff

☞ Know your project

☞ Lead by example

☞ Lead from the front

☞ Organize and develop the staff

☞ Organize and communicate specific tasks and responsibilities

☞ Demonstrate and generate enthusiasm for the project

Charge out rates for staff

Probably nothing generates more confusion, distrust and envy than the matter of charge out rates. Usually either the client believes that the consultancy is making an abnormally high profit from the use of contractual staff, or the users believe that the consultant is in receipt of this abnormally high remuneration.

Care must be taken to ensure that the charge out rates for your staff accurately reflect the costs of undertaking the work, but also great care must be taken to ensure that the clients are aware precisely what they are going to receive for their expenditure.

Charge out rates are usually the only method of the consultancy receiving payment for work done. The exception being a fixed price contract, but in this case, the Project Manager must be more aware of the true costs of his staff.

For the Project Manager and senior consultants, much pre-contract work is undertaken speculatively and at no cost to the client. This "free time" must also be computed into the equation.

As a rough guide, 50 % of the costs should appertain to the remuneration package. Many large companies have charge out rates of 2 - 3 times the team member's remuneration. This is acceptable and is due to the heavy overheads of large company and the costs of the support functions that allow the project to be gained and funded.

The charge out rate for a member of the project team should be calculated as follows.

Table 3 Charge Out Rates

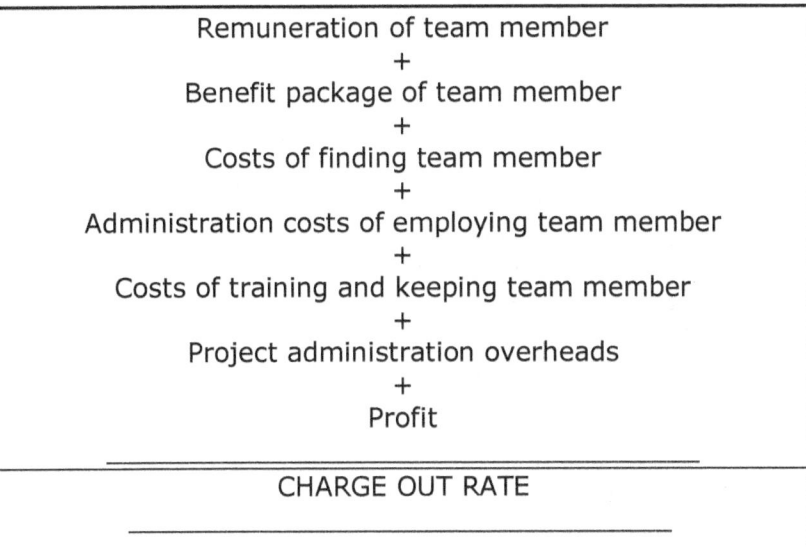

Due care should be taken to ensure opportunity costs - the income lost from not pursuing other activities, are also computed within the equation.

Choosing your team member

Having decided upon the constituents of the project team and the amount of money that can be charged for their services, you now need to choose and employ the staff.

The first thing that needs to be done is the completion of a Terms of Reference (TOR) for each position. This would consist of the following:
- Job title
- Description of working environment
- Technical architecture
- Job description
- Essential skills
- Desirable skills
- Terms and conditions
- Application details

The resulting applications should be reviewed and interesting applicants short listed for interview.

It is seldom fruitful to interview more than four people for any one position. This is due to time and cost constraints.

There are many different ways of interviewing but it is important that the interviews are consistent, pertinent and fair for all. The object is to find the appropriate person for the job, but sometimes that person may not be available so sometimes the best candidates should be chosen. Do not be afraid to re-advertise the position - the costs to your reputation and the project can be great of an inappropriate or experienced person.

The following are the usual factors to be examined at an interview:
- Presentability
- Personality
- Experience
- Qualifications
- Ability
- How they would interact with the rest of the team
- Their suitability for the job

The following is a suggested interview format:

Outline of the available position	Interviewer 10 %
Their aptitude for the position	Interviewee 20%
Set questions to ascertain experience etc.	Interviewer 20 %
Demonstration of experience and capabilities	Interviewee 40%

Project Stages

Project Methodologies: Different Project Managers and consultants follow different methodologies. A methodology is a structured route to guide the project team through a project life cycle. It defines the staffing structure and responsibilities of the team members, the tasks that have to taken and the deliverables from each stage of a project.

Most of the methodologies define four main areas of project responsibilities - client (and/or sponsor), business (and/or users), information technology (and/or project team) and quality.

The major methodologies have a common core of types of work to be undertaken. They vary in the business/technology emphasis, the types of reports required and the titles of these reports and stages. They all have a common feature of dividing projects into stages, each having a given order and clearly defined deliverables (or exit criteria) for each stage.

Why Divide a Project into Stages?

As all the major methodologies divide projects into stages it is tempting to ask why?

To ensure:
- Better control - the project is divided into small manageable sectors, each with a clearly defined group of actions and deliverables that form - baselines on which to build the next stage.
- Easily identified and corrected errors by the use of a clearly defined route map and requirement, which highlight any short fall.
- All required actions are taken in the correct order by following the route map.
- Appropriate resource allocation in the correct order. The defined stages and route map indicate when in the project different types of resources are required.
- A firm foundation to the project plan. Adherence to the methodology produces an outline project plan.

FastTrack © Project Management

To facilitate:
- Easily understood progress reporting using the standard documentation.
- A series of attainable goals - every one can follow the route map of what should happen when and what deliverables are available.
- Easier planning and reviews - which are built into the methodology.
- Effective work scheduling by the use of a standardized project plan.
- Staged reviews, sign off points and payments that are understood by all.

The Main Project Stages

Whilst the terminology may differ amongst methodologies the following stages should all appear. In some methodologies two or more stages may be grouped together.
- Project Inception - the start up of the project where staffing, budgets and resources are made ready. This stage usually involves a skeleton staff. There may also be some pre-sales activity still being undertaken.
- Feasibility - one or more technically and/or business skilled consultants review the requirements and possible solutions and recommend one or more feasible options.
- Project Planning - planning how to approach the project, obtaining staff and resources.
- Quality Assurance - happens throughout the project and is the measurement of outputs against agreed quality standards.
- Analysis - the investigation of existing systems and how to define the proposed system.
- Design - of the new system.
- Development of the new system.
- Testing of the new system.
- Implementation of the new system.
- Data Conversion and Take on converting data from the old system and transferring it to the new.
- System Support of the new system.
- User Support and Training.
- User Acceptance.

Project Reports

In some small projects some of the reports in the analysis stages may be consolidated either with each other, or with the system or support documentation.

Whilst some of the reports may appear to hold similar contents, this is not duplication. Either the recipients may be different or the contents need to be enhanced or strengthened during the life of the project. Some reports may be sub reports, designed for a particular purpose or audience.

Reports can be described as "living" - subject to continual but controlled update (e.g. project plan) or "baseline" - unchanged except by formal change control (e.g. User Requirement Specification)

More detailed explanations or the contents and recipients of the reports are given under the succeeding sections.

Project Inception

Aim:
- Establish the existing problem or the business requirement that the system must resolve.
- Obtain preliminary definition of aims of proposed system.
- Establish outline of resources, time and budget available.
- Obtain formal sponsorship for proposed project at a senior level

Actions:
- Interviews, discussions and consultations with proposed sponsors, senior users and budget providers.
- Prepare Terms of Reference with respect to role of project team and consultants.
- Initial macro analysis of the site, existing system, problems and solutions presently being considered.
- Initial macro level discussions with senior users.
- Outline walk through and outline document flow diagram of existing system
- Obtain outline costs and staffing guesstimates (± 30%)

- Prepare Statement of Scope and Objectives to include outlines of proposed system, resources, time and budget available. This is often included within the Project Charter, but sometimes the Statement needs to be approved by another management group or prior to the formal presentation of the Project Charter
- Prepare Project Charter, giving only outline details of the Cost Benefit Analysis, Risk Analysis and Feasibility.
- Prepare, present and obtain formal sign off of documentation.
- Obtain authorization of Project Charter to enable project to proceed to next stage.

Documents Produced:
- Terms of Reference (TOR)
    - Executive Summary
    - Description of work to be undertaken by consultants and project team.
    - Time and budget allocation
    - Sign off by Sponsor

- Statement of Scope and Objectives:
    - Business Requirement
    - Executive Summary
    - Project Scope
    - Project Objectives
    - Project Constraints
    - Sign off by Sponsor

- Project Charter:
    - Project Sponsor
    - Business Requirement
    - Executive Summary
    - Definition of project
    - Scope of Project
    - Project Objectives
    - Business Case
    - Feasibility
    - Identified Time Constraints
    - Ramifications of project
    - Cost Benefit Analysis
    - Risk Analysis

- Recommendations
- Steering Committee Authorization

Feasibility

Aim:
- To define the scope and objectives of the proposed system.
- To measure and weigh the benefits against the risks and costs of proceeding with the project.
- To assess whether or not to proceed with a particular project before any major planning or expenditure is undertaken.
- To assess the viability of the system with respect to the following:
    - Will it meet the Business Requirement?
    - Is the risk of undertaking the project acceptable?
    - Is the required system and technical architecture appropriate to present technology?
    - Are the proposed budget and resources appropriate?
    - Are the costs of the system and its operating costs justified?
    - Will the system meet the user's expectations and requirements?
    - Will the system save the clients money and make them more profitable?
    - Will the system improve productivity and efficiency?
    - Will the system provide better management information allowing for better decisions to be taken?

Actions:
- Interviews, discussions and consultations with proposed sponsors, senior users and budget providers.
- Prepare Terms of Reference with respect to role of project team.
- Outline macro analysis of the site, existing system, problems and solutions presently being considered.
- Outline macro level discussions with senior users.

FastTrack © Project Management

- Macro walk through and outline document flow diagram of existing system
- Obtain outline costs guesstimates (± 30%)
- Define at least four options, including "no action" to meet the Business Requirement
- Ascertain and measure the risks associated with each option
- Ascertain and measure the benefits associated with each option. Ensure that tangible and intangible benefits are documented
- Ascertain and measure the costs associated with each option. Ensure that tangible and intangible costs are documented.
- Prepare a cost benefit analysis (CBA) from the above three actions.
- Prepare, present and obtain formal sign off of documentation.
- Undertake a SWOT (Strengths, Weaknesses, Opportunities, Threats) analysis on the system, proposals and Business Requirement.
- Ascertain any constraints that may be placed upon the project or the system.
- Obtain sign off to proceed to next stage.

> ⚑ Note: at this stage you are investigating the feasibility of the proposed system and NOT producing an outline system design.

Documents Produced:
- Project Initiation Document (PID):
    - Project Sponsor
    - Executive Summary (background, reasons, aims and constraints)
    - Business Requirement
    - Terms of Reference
    - Project Scope and Objectives
    - Business Case
    - Ramifications of Project
    - Feasibility
    - Findings from the SWOT analysis
    - Risk and Cost Benefit Analysis
    - Review of Possible Solutions (at least 4 including "no action")

- Costs and Benefits of each Solution
- Recommended Solution
- Project Organization and Responsibilities
- Proposed Schedule
- Sign off by Sponsor

Project Planning

Aim:
To ensure that:
- All the baselines and resource allocations have been properly ascertained, identified, documented and agreed.
- The project is sectioned into stages, tranches and the various milestones and deliverables have been identified.
- The various tasks within the project have been fully identified and documented in the correct sequence, and the appropriate time and resources allocated to each of them.
- All the dependencies of one task or resource allocation upon each other, have been identified and the appropriate actions taken.
- The clients, users and project team all have detailed knowledge of the project progress and future requirements and work.
- All the required resources are allocated to the project at the appropriate time.
- Risk and costs are minimized whilst resource usage, quality and system production is maximized.
- Reliable delivery dates for each milestone and deliverables are available and known to all.
- Changes and progress to the original project plan is easily made and documented.
- Reliable reporting and control documentation is readily available on resources, quality and risk.

Actions:
- Define the deliverables, project stages, milestones, major activities, tasks, dependencies and activity time frames.
- Ascertain resources and resource availability over the project time frame.
- Define the Sponsor and the Business Requirement

FastTrack © Project Management

- Write the Executive summary, Ramifications / Impact on client environment, Proposed schedule, Scope and Objectives
- List all activities and tasks in a logical format, including identifying Project tranches and any sub-projects.
- Allocate dependencies and resources to each.
- Re-list all activities and tasks until a workable format is found, using an agreed project plan format.
- Document and re-plan as project progresses or change control is incepted.
- Write the Project Plan
- Review it with management
- Gain client Sign Off to proceed to User Requirements
- Write the Project Plan (Requirements) based on analysis with client
- During requirements specification present Project Review for approval.
- Update Project Plan with
  - Milestones
  - Costs analysis ($\pm$ 30%)
  - Actions completed
  - Actions still outstanding with new estimated completion date
- Re-present Project Plan to client regularly or as major changes made.
- Gain client Sign Off to proceed with changes via change control.

The Project Plan should be a living document and updated regularly from contributions from all the project team.

Documents Produced:
- Project Plan
  - Project Name
  - Project Sponsor
  - Business Requirement
  - Major Deliverables and Milestones
  - Project Stages
  - Major Dependencies and Constraints
  - Project Plan
  - Gantt, Pert as required
  - Work Assignments
  - Project Reviews

FastTrack © Project Management

- Staffing Plan
    - Staff Tree
    - Staff Terms of Reference
    - Numbers and time allocation of staff
    - Costs of staff by staff type and period
    - Proposed staff budget
    - Details of any staff proposed.
    - User staff involvement required.
    - Updates
    - Authorization

- Budget
    - Budget with respect to period, group and /or phase
    - Variances
    - Reasons for Variances
    - Estimated future spend
    - Authorization

- Project Review Report (staff)
    - Name
    - Section
    - Achievements during period
    - Meetings and movements
    - New work assigned but not scoped
    - For management attention
    - Time sheets
    - Outstanding issues
    - Plans for next week
    - Meetings and movements next week
    - To Do List
    - Signature

- Project Progress Report (users)
    - Project Name
    - Achievements during period
    - Problems encountered or anticipated
    - Planned activity for coming period
    - For management attention
    - Project Manager's Authorization

Quality Assurance

Aim:
- To ensure that the project team, the clients and the users are aware of what is required to produce a good quality system.
- To define the quality reviews and audit mechanisms, controls and reports.
- To define the procedure for rectifying matters that fail the quality reviews
- To ensure that the project and system adheres to the appropriate quality standards e.g. ISO9000

Actions:
- Ascertain and define the requirements of the quality report
- Set up the quality review procedures
- Ensure that all concerned are aware of the quality requirements by the use of reviews, marketing, workshops, training and reports.
- Feed the requirements into the project plan
- .Update as reviews, progress and change controls dictates.

Documents Produced:
- Quality Plan
  - Project Name
  - Project Manager
  - Quality Manager
  - Project Standards
    - Analysis and design methodology
    - Documentation
    - Coding
    - Testing
  - Deliverables
    - System
    - Programs
    - Test specifications
    - Documents
    - Hardware
    - Operating software
  - Schedule and independent audits
  - Testing procedures including acceptance test project plan.

Analysis

Aim:
- To become fully conversant with the Business Requirement
- To fully document the existing system
- To ensure that the requirements of the users can be defined and specified in a format that both the clients and the technical project staff can understand.
- To prepare the basis for the design to the new system.
- To prepare and have agreed the baseline User Requirement Specification (URS), against which all Changes are measured.
- To have an agreed definition of the client's requirements, proposals and ideas about the new system.

Actions:
- Investigate and document present functionality, data and system users
- Undertake full analysis of all existing data, its uses, breakdown and actions upon it.
- Identify technical platform
- Produce formal diagram of data flow and document flow
- Undertake a full analysis of all costs appertaining to the project, and update project plan with these details
- Produce an User Requirements Specification (URS) for each tranche
- Present the URS and obtain the client's acceptance

Documents Produced:
- User Questionnaires
    - Project Phase
    - Activity, Data or Action
    - Questions and either free flow or choice of answers
    - Collator of information
    - Time frame

- User Requirement Specification (URS)
    - Executive Summary
    - Computer Strategy
    - Business Constraints and Dependencies
    - System Objectives
    - Existing Workflow
    - Proposed Workflow
    - Specific System Requirements
        - Technical Platform
        - Hardware
        - Software
        - Functionality
        - Look and feel
        - Structure
        - Interfaces
        - Dependencies
        - Comms
        - Maintenance and Support
        - Housekeeping and Audit
        - Security
        - Environment
        - Data transfer
        - Required Benchmarks
    - People Based Requirements
        - Procedures and Methods
        - Users Involvement
        - Training and Technology Transfer
    - Organizational Based
        - Structural and Organizational
        - Jobs
    - Summary of Problems
    - User Authorization

- Functional Specification
    - A technical specification and description of the proposed system, produced for the benefit of the systems analyst.

- Document Flow Diagrams
    - A diagrammatic description of each document and how it moves (flows) through the existing and proposed system. Details of changes enacted upon each document. Produced for the systems analyst.

- Data Flow Diagrams (DFD)
  - Similar to the Document Flow Diagram except that the flow of each piece of data appertaining to the documents and the system is described. Produced for the systems analyst.
- Entity Relationship Diagrams (ERD)
  - A diagrammatic description of each small part (entity) of the system interacts with other parts of the system. Produced for the systems analyst.
- Entity Life History (ELH)
  - A diagrammatic description of each small part (entity) of the system interacts and changes over the life span of the system. Produced for the systems analyst.

Design
Aim:
- To take the detailed analysis of the business analyst and the consultants and produce a technical design of the required software.
- To confirm the hardware and communication configurations required for the software.

Actions:
- Define database structure
- Document system hierarchy
- Define Screen and Report layouts (or prototype)
- Design program units
- Design required interfaces
- Document inter-dependencies
- Document benchmarks
- Produce System Specification (for each tranche):
- Define system test plan
- Walkthrough the design
- Update Project Plan with
  - Milestones
  - Costs analysis (± 10%)
  - Actions completed
  - Actions still outstanding with new estimated completion date

- Re-Present Project Plan to client
- Present System Specification to client
- Gain client Sign Off to proceed to Program Development

Documents Produced:
- System Specification
  - System Overview
  - Technical Architecture
  - Data Flow
  - Database Structure
  - System Hierarchy
  - System Inter-dependencies
  - System Interfaces
  - Benchmarks
  - Authorization

Development

Aim:
- To develop code and test the software programs as specified in the analysis and design stages.

Actions:
- System Development for each tranche or module:
  - Develop program units
  - Update Data Dictionary
  - Test program units
  - Document within the system programming
  - Produce Help Screens
  - Produce Program Specification
  - Integrate system components
  - Test system
  - Error correction
  - Produce User and System documentation
  - Release and control software versions, using Version Control
  - Undertake any Change Control requirements.
- Gain client Sign Off to proceed to implementation

Documents Produced:
- Program Specification
  - Overview
  - Data structure
  - Program structure
  - Program "Called by"
  - Program "Calls"
  - Files Used
  - Screen standards
  - Functional Description
  - Benchmarks

- System Documentation
  - System Overview
  - System Hierarchy
  - Units in System
  - Files in System
  - Appendices

- Data Dictionary
  - Details of all the variables and fields used within the program, modern RDBMS's produce the data dictionary automatically.

- Version Control
  - Version Control Procedures
  - Release numbering
  - Deliverables schedule
  - Version isolation procedures

- Error Report
  - Project Phase and Task
  - Date and Log Number
  - Priority
  - Instigator
  - Error
  - Status
  - Requirements
  - Estimated Costs
  - Completion sign off

Testing

Aim:
- To define the types of tests, methods and tester's responsibilities
- To fully and formally test the software.

Actions:
- Define testing prerequisites
- Define the test structure and scope, the expected method of testing and expected results, include any special instructions.
- Describe the test schedule of the expected effort to conduct each test module, the order of testing, and the testing assignments will be included in the test plan.
- Undertake Unit Testing, usually conducted by the programmer responsible for production of the unit under test; generally the person who has programmed the unit. The purpose of unit testing is to qualify individual components of a system. The tester ensures that the unit meets the program specification
- Document the testing performed
- Make any corrections necessary, notifying the Project Manager as errors are corrected.
- Undertake System Testing, based on the Test Plan which is prepared in the Design phase of the project life cycle. The purpose of system testing is to qualify the system as a whole, by testing the menu structures, user and program interfaces, etc... The tester may be anyone who has not programmed the part of the system under test.
- Compare the expected and actual results, and if within limits, that system part will be deemed to have passed.
- Document the testing performed.
- Pass the test reports to the Project Manager for review.
- Make any corrections necessary, notifying the Project Manager as errors are corrected.

- Undertake User Acceptance Testing, based on the Test Plan that is prepared in the Design phase of the project life cycle. The purpose of user acceptance testing is to qualify the system as a whole, to the satisfaction of the Project Sponsor, ensuring that the Project Objectives have been met and the User Requirements satisfied.
- Document the testing performed.
- Pass the test reports to the Project Manager for review.
- Make any corrections necessary, notifying the Project Manager as errors are corrected.

Documents Produced:
- Test Plan
    - Objectives and standards to be met
    - Types of tests
    - Source of test data
    - Responsibilities
    - Testing Plan
    - Control procedures
    - Quality Manger's Authorization
    - Project Manager's Authorization

- Test Acceptance Report
    - List of Tests carried out
    - Review of testing activities
    - Test data used
    - Results
    - List of outstanding problems
    - Recommendations
    - Programmer 's or testers sign off
    - Quality Manager's authorization
    - Project Manager 's authorization

FastTrack © Project Management

Implementation

Aim:
- To fully install the tested technical platform, peripherals and software within the client's site.

Actions:
- Undertake any site preparation necessary
- Ensure that site is prepared for installation and implementation
- Install system in test environment
- Configure and install the hardware, software and network.
- Install software in the production environment (client's site)
- Undertake any error corrections necessary.
- Act upon any changes requested by the client, using the formal Change Control procedure
- Undertake a Quality Assurance review
- Provide user department with system training
- Present system to user for User Acceptance Testing

Documents Produced:
- Release Control
  - List of Contents
  - Version Identification
  - Description of new and changed functions
  - List of Error Reports fixed in the release
  - Technical Architecture
  - System overview
  - Deliverables schedule identifying each object in the release.
  - Installation or Upgrade Guide
  - Training material updates
  - User and Operations Manual updates
  - Help desk or support function details
  - User Acceptance sign off

- Change Control
  - Project Phase or System
  - Date
  - Log Number
  - Instigator
  - Priority

- Description of required change
- User Authorization
- Change Definition
- Project Manger Authorization
- Project Control Office Check
- Allocation
- Project Plan for the change
- Estimated costs
- Impact of the change
- Project Manger Authorization
- User authorization
- Project Control Office Check
- Work Allocation
- Allocation acknowledgment
- Work completed
- Cost , time and impact notification
- Quality Assurance
- Testing
- Project Manger Authorization for implementation
- User authorization
- Project Control Office Check
- Installation Confirmation
- User Acceptance

Data Conversion and Take on

Aim:
- This stage is not always necessary as many system are implemented with data available from the implementation date onwards
- Where there is a wish to transfer data from the old system to the new, the aim would be to ensure that the data is transferred in a timely, accurate and cost effective manner.
- Data transfer is often treated as a separate, but linked, project from the main project. The alternative is to treat it as a sub-project.

Actions:
- Ascertain feasibility and necessitation of transferring data from the old system to the new.
- Ascertain costs, resources and time required.
- Prepare project plan and staffing schedules
- Undertake transfer
- Undertake period of parallel running
- Test new data against a known test pack of expected results
- Undertake User Testing
- Obtain User Acceptance

Documents Produced:
- The reports required are those of Feasibility, Project Planning, Implementation and Testing, written as appropriate to the data transfer.
- Documentation and Support is usually be encompassed within the main project.

System Support

Aim:
- To ensure that the system is fully documented, such that future enhancements can easily be made.
- To ensure that the client's technical staff are able to support and maintain the system.

Actions:
Draft, review and adopt formal procedures for the day to day running of the system.
Draft, review and adopt formal procedures for any future enhancements to the system
Write System Documentation
Write User Support Documentation

Documents Produced:
Procedure Documentation
- System overview
- Procedure overview
- Flow chart of procedures
- Contact for Help

System Documentation
- System overview
  - Role of the system within the operational environment.
  - The functionality of the system.
- System hierarchy
- Units in system
- Files in system
- Appendices

– User Support Documentation
- System overview
- Installation
- System hierarchy
- Operational guide by function
- Occasional maintenance
- Configuration
- Help Desk details

User Support and Training
Aim:
– To ensure that the Users are able to use, support and maintain the resulting system.

Actions:
– Create User Documentation
– Provide client's support group with system training
– Provide User department with system training
– Conduct user training
– Conduct Hardware, software and network training
– Formally advise clients of support and maintenance available from suppliers and/or consultancy
– Revisit training as required
– Update documentation as required under any support contract
– Formally transfer to support and maintenance contract as appropriate

Documents Produced:
- Help Screens
- User Handbook - quick guide to common procedures and problems
- User Guide
- Procedures Guide
- Operations Guide
- Management Overview
- Installation and Maintenance Manual

User Acceptance

Aim:
To ensure that the system:
- Meets the Business Requirement
- Meets the User Requirement Specification
- Meets the acceptance criteria
- Meets all the benchmarks
- Is fully and completely documented
- Is of sufficient capacity
- Has fully tested and acceptable audit, maintenance, back-up and recovery abilities
- To obtain a formal sign off to the project ☺
- To facilitate the payment of the consultants ☺
- for the final Deliverable

Actions
- Conclude User Acceptance Testing
- Walk through and demonstrate the system
- Undertake a period of parallel running of the old system against the new.
- Market and promote the new system, as required.
- Ensure that training is underway and satisfactory
- Ensure that documentation is complete
- Ensure that all quality checks are completed
- If the Project Sponsor is satisfied with the system, invite him to sign-off the project.
- The sign-off will constitute the formal hand over of the system to the User department.

Documents Produced:
- User Acceptance Test Report
  - Project Name
  - Deliverable
  - Management Overview
  - Review of testing activities
  - Test data used
  - Results
  - Quality Manager's authorization
  - Project Manager Authorization

- User Acceptance
  - Project Name
  - Overview of Aims of project
  - Review of costs and time scale of project against estimated and projected figures
  - Overview of problems encountered
  - Overview of changes made
  - Any future proposals
  - Formal sign off of Project Manager
  - Ramifications of formal acceptance
  - Steering Committee authorization
  - Invoice

Table 4 Summary of Documents Produced

Stage	Reports	Activity
Project Inception	Terms of Reference (TOR)	Plan, Communicate
	Statement of Scope and Objectives	Communicate
	Project Charter	Communicate
Feasibility	Project Initiation Document	Control
Project Planning	Project Plan	Plan
	Staffing Plan	Plan
	Budget	Plan
	Project Review Report (staff)	Plan
	Project Progress Report (user)	Communicate
Quality Assurance	Quality Plan	Plan
Analysis	User Questionnaires	Communicate
	User Requirement Specification (URS)	Control
	Functional Specification	Control
	Document Flow Diagram	Document
	Data Flow Diagram (DFD)	Document
	Entity Relationship Diagram (ERD)	Document
	Entity Life History (ELH)	Document
Design	System Specification	Control
Development	Program Specification	Control
	Program Documentation	Document
	Data Dictionary	Document
	Version Control	Control
	Error Report	Control
Testing	Test Plan	Plan
	Test Acceptance Report	Communicate
Implementation	Release Control	Control
	Change Control	Control
Data Conversion	stages as for standard project	

FastTrack © Project Management

Stage	Reports	Activity
System Support	Procedure Documentation	Document
	System Documentation	Document
	User Support Documentation	Document
User Support	Help Screens	Document
	User Handbook	Document
	User Guide	Document
	Procedures Guide	Document
	Operations Guide	Document
	Management Overview	Document
	Installation and Maintenance Manual	Document
User Acceptance	User Acceptance Test Report	Communicate
	User Acceptance Sign Off	Control

Table 5 Generic Project Plan

Project Phase	Task Description
CONSULTANCY	Client relationships
	Project Management
	Contract negotiations
	Contract definition
	External supplier selection
	Tendering
	Supplier relationships
ADMINISTRATION	General administration
	Staff training
	Travel
	General support
	Staffing Issues
	Project meetings
	Management meetings
PROJECT INCEPTION	Business requirement definition
	Terms of Reference
	Statement of Scope and Objectives
	Project Charter
	Project Charter Presentation & Acceptance
FEASIBILITY	Project definition
	Cost Benefit Analysis
	Risk analysis
	SWOT analysis
	Feasibility study
	Terms of Reference - project team
	Project Initiation Document
PROJECT PLANNING	Project Plan
	Staffing Plan
	Budget
	Executive Summary
	Project Review
	Project Report
QUALITY ASSURANCE	Quality Plan
	Quality review

FastTrack © Project Management

Project Phase	Task Description
ANALYSIS	Data analysis
	System analysis
	Costs analysis
	Logical data design
	Technical platform
	User Questionnaire
	Functional Specification
	User Requirements Specification
	URS Presentation & acceptance

Project Phase	Task Description
DESIGN	Database design
	Systems design
	Program structuring
	Interface design
	Program design
	Screen & report prototypes
	Design walk-through
	Systems Specification presentation & acceptance
DEVELOPMENT	Unit programming
	Unit testing
	System integration
	System testing
	Error correction
	Benchmark testing
	Program specification
	Program documentation
	Release control
	Change control
	Version Control
	System documentation
	User documentation

Project Phase	Task Description
TESTING	Test Plan design - system
	Test Plan design - user acceptance
	Test schedule
	Test environment installation
	System testing
	Test Acceptance Report - system
	User tester training
	User acceptance testing
	Error correction
	Test Acceptance Report - user acceptance
IMPLEMENTATION	Site preparation
	Hardware/software/network configuration
	Hardware/software/network installation
	Change control
	Release Control
	Quality Assurance
SYSTEM SUPPORT	Procedure drafting
	Procedure review
	Procedure adoption
	Procedure Documentation
	System Documentation
	User Support Documentation

FastTrack © Project Management

Project Phase	Task Description
USER SUPPORT & TRAINING	Hardware/software/network training
	User training
	Management Overview
	Operations Guide
	Procedures Guide
	User Guide
	Help Screens
	User handbook
	Installation and Maintenance Manual
	Formally advise clients of available support and maintenance
	Formal transfer to support and maintenance contract
USER ACCEPTANCE	System walk-through
	Quality check
	Market system
	Promote system
	Acceptance Test Report,
	Formal handover
	Sign-off from Project Sponsor
	User Acceptance
DATA CONVERSION	Feasibility
	Transfer
	Parallel running
	Test pack testing
	User Testing
	User Acceptance

Project Planning

Basics of a Project Plan: A project plan is a three dimensional entity of Time, Task and Resource plotted against each other, with each of the axis comprising of several factors. Thus the production of a project plan is both complex and time consuming. Thankfully there now exists, several software packages that assist with the production of the plans. The model below explains how Time, Task and Resource are linked together.

Table 6 Project Planning Model

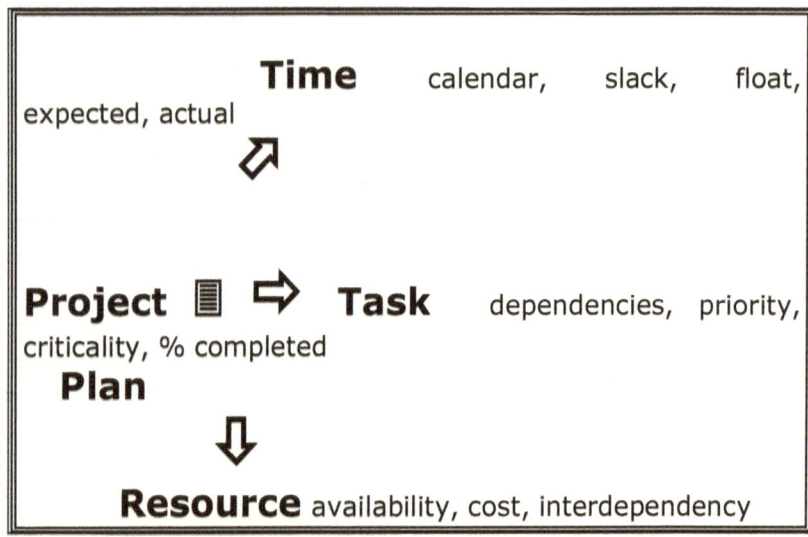

There is no alternative to the construction of the plan but by the ascertainment and input of much basic information coupled with sufficient knowledge of how to define the project elements. Experience will teach the basic structure of a plan, and many Project Managers have in their "toolboxes" a number of standard project plan layouts. Below are a number of standard definitions of the components of a good project plan.

Budget - The amount of money allocated to a project. It is usually separated into amounts per resource or action per period.

Baseline - A formally defined, agreed and non moveable or changeable starting point. It is used to measure variances in quality, risk, budget, resources etc. Also used when investigating problems within a project. The baseline is the known entity. The baseline is also used when releasing a sub-system or system, so that a know version can be released and all other changes and enhancements can be released as subsequent versions.

Milestones - Major events or deliverables within the project plan.

Task - An action, with a well defined entity that must be undertaken in order to achieve or produce something. Tasks are often logically linked by action or time. Tasks can be assigned a priority or criticality depending upon how important it is.

Macro Task - A group of tasks that logically link together - used in planning and for ease of reference

Critical Path - the absolute, must do tasks on the project plan. These show the path that it is critical to follow. The critical path is defined by the project management software and usually consists of the shortest route from beginning to end of the project. All of the tasks must be completed on time in order not to adversely affect other tasks. The tasks on the critical path are the lowest level of the dependent tasks e.g. the one that one or more tasks are dependent upon.

Deliverables - A tangible, significant part of the resulting system, the implementation of which marks a milestone within the project plan. Major Deliverables are often part of the contract between the project team and the users. All deliverables should feature within the project plan. An example is a major report, a software module or a computer sub-system.

Dependency - A constraint on the sequence and timing of a task within the project plan. When writing a realistic project plan it is important to take into account dependencies between tasks. Most project management tools allow the user to define the dependencies and then will automatically define the critical path for the project. Dependencies have one of the following characteristics:
- Finish/Start - the start of one task is dependent on the finish of another
- Start/Start - the start of one task is dependent on the start of another
- Finish/Finish - the finish of one task is dependent on the finish of another
- Lag/Lead - there is a defined time lag or lead time between the start of one task and the start of another
- Slack - the amount of time between one task and another
- Float - the total amount of "spare" time in a project plan e.g. the culmination of all the slack times. Experienced planners build some float into the plan to take in account the unexpected situations.

Phase - A division of the project plan, which denotes a group of complimentary project tasks - e.g. training

Stage - Part of a project, denoting a group of complimentary project tasks, that has a formal, identifiable beginning and ending and is marked by a major deliverable - e.g. testing

Why Project Plan

To ensure that:
- New systems are delivered on time, within budget and of good quality.
- Inevitable problems are recognized, planned around and encapsulated within the work done.
- A firm foundation for effective systems development is provided.
- Slow and methodical planning at the beginning of the project is better than frantic control and fire fighting at the end of the project.
- All areas of the project have the same methodical attention to their inception - so removing possible problem areas.
- Deliverable dates and end dates can be accurately predicted and met.
- Budgets, manpower requirements and resource needs can be predicted and efficiently utilized.
- Work schedules are available and purposeful.

Problems of Project Planning

The following is a good example of common problems experienced when trying to plan a project:
- It's a necessary evil that people generally dislike doing
- Unrealistic targets being set by the clients - either from enthusiasm, ignorance or political needs
- Unrealistic budgets being set, or realistic budgets not being available
- Much of project planning relies on experience
- The sheer complexity of the process, taking into account so many unknown variables
- Need for constant but controlled re-planning and reviewing
- The necessary project administration needs dedicated team members to administrate and control it.
- Care must be taken to ensure that just enough planning and updating is undertaken to be effective without the volume of work and paper being counter-productive

Planning Tools

Gantt Chart - A graphical way of project planning- shows actions over time-bars.

Pert Chart - A project planning technique, concentrating on time that employs statistical analysis to estimate the probability of meeting target dates.

Work Breakdown Schedule - A graphical way of showing the hierarchical relationships between tasks and headings, produced by many planning software packages

Time and Work Allocation - A document produced by many planning software packages, or independently, that outlines each team member's tasks to be undertaken and the time frame assigned to this work.

Flow Charts - a diagram consisting of boxes and linkages that represent the flow of work over time and distribution. It is used as an analysis tool and to outline project task flows.

To-Do Lists - A prioritise list of tasks in hand. Useful as a personal task list, originating from the more micro view of the project plan.

Walkthroughs - A "walk" by the Project Manager of the individual steps that the project will follow - can only be used for small project or at macro level. Often accompanied by a flow chart and used either in the rough planning stage, or to check a completed project plan. Can also be used to demonstrate relevant parts of the plan to clients, users or team members.

Steps to a Project Plan

The following is the usual route to take when setting up a project plan using a planning tool:
1. Project Details - the description, budget, start and finish date, defaults and project calendar
2. Staff Calendar - standard day, charges, staff availability
3. Staff Details - staff groups, involved users, efficiency
4. Resource Details - resource groups, availability, costs

FastTrack © Project Management

5. Project Specification - rough paper outline of description, deliverables and resources
6. 6) Divide project into Phases - consider the use of several linked, interdependent project plans
7. Ascertain Project Stages - these are for planning clarification and have a duration of zero
8. Define the Deliverables - allocate the delivery date as fixed date
9. Define the Milestones - these are for planning clarification and have a duration of zero
10. Define any other known and agreed actions with fixed dates
11. Input standard tasks
12. Input other tasks
13. Define and input task dependencies
14. Ascertain, using the software, check and correct the critical path
15. Ascertain, using the software, check and correct the Work Breakdown Structure (WBS)
16. Ascertain, using the software, check and correct the continuation of linkages
17. Produce a printout the required reports -Gantt, Time and Work Allocation, Work Breakdown Schedule, Budgets, Resources

Steps 11 - 13 are iterative. On the first run through - it is normal for the activities merely to be listed with no time frames. The inputting of the known and agreed fixed dates and the dependencies will allow the software to assist with the process. Task time frames must be a factor of budget, resource availability, dependencies, efficiency and time it takes to undertake the work. All except the last factor can be ascertained with the assistance of the software.

Try to work forwards - i.e. from start to unknown finish date, as opposed to backwards - from set finish date to start date, concertinaing tasks between the two dates.

Don't forget to include user activities within the plan - particularly if a project team action is dependent upon the task being completed.

Transition Planning

This is planning the movement from one computer system to another. The following will need to be considered.
- Identify the impact and implications of the system development
- Define the deviations from normal routine
- Plan transition to produce minimum disruption
- Define people involved:
  - Management
  - Users working with design and development team
- Ascertain time and time frames of people required
- Define time needed for the new site preparation
  - Hardware and comms installation
  - Operating system installation
  - Software installation
- Define time needed for user and project team training
- A period of parallel running will be needed; a common period is one month.
- Time will be needed to collect and act upon user feedback
- Arrange for user availability
- Arrange for back up resources
- Cost and prepare budgets
- Update project plan

Budgets

Budgets, as with all of project planning is a continuing process, where controlled changes to the baselines are being continually made, as the project evolves.

Budgets can be broken down into manpower and computing resources - both of which can be planned and controlled within most standard planning software packages.

Project budgets are usually ascertained from one of the following scenarios:
- The budget is fixed and the time frame must be ascertained from the effective use of this budget limit.
- The project time limit is fixed and the budget must be ascertained from the most efficient use of resources to achieve the required deliverables within this time frame
- The resource availability and unit cost is fixed. The time frame and total budget must be ascertained by the effective trade-off of time and resource usage in order to achieve the required deliverables.

It must be emphasised that, although the total budget is the responsibility of the Project Manager, the whole project team has a responsibility for budget usage. This factor is best controlled by the issue of properly controlled, sub-budgets to team leaders, or responsible team members.

The setting of budgets should be iterative, the more detailed the requirements specification and project planning, and the more accurate the budgets can be. Thus at Project Inception budgets on a small to medium project can be accurate to ± 40 - 50%, whilst after feasibility an accuracy of ± 10% would be anticipated.

Contingencies to budgets are always added. These take the form of a percentage uplift. Many Project Managers clearly mark this figure in their budgets. Unfortunately Client's and team members often misinterpret this figure to be Consultancy profit or money to be spent whatever the circumstances. To overcome these problems, many Project Mangers also build in slack to many of their more risky costings.

Profits are built into the project by means to use of charge out rates on staff and consultancy resources and the costs of materials and equipment sold through the consultancy.

Project overheads, such as marketing, contract negotiation and acquisition, administration staff and costs are usually shown in the budget as a percentage uplift or a standard daily charge out rate.

However costs, contingencies and profits are calculated, the whole budget must be reasonable and defensible. Hiding costs and profits within a budget, not only makes it difficult to allocate and control, but is also unprofessional and open to criticism and abuse.

Steps to Planning a Manpower Budget

The following is a suggested method of planning a manpower budget:
- Ascertain project team members, their Terms of Reference (TOR), charge out rate and their skills
- Ascertain project tasks from the project plan *
- Allocate team member (s) to each project task *
- Calculate the total cost for each team member or team group by phase and total project. *
- Adjust team allocations and task duration until budget and project plan are realistic, feasible and efficient. *

Alternatively:
- Ascertain project team members, their Terms of Reference (TOR), charge out rate and their skills
- Define Time and Work Allocation schedule *
- Allocate project members to each work allocation *
- Calculate the total cost for each team member or team group by phase and total project. *
- Adjust team allocations and task duration until budget and project plan are realistic, feasible and efficient. *

This is usually done via the project planning software

An Example of a Budget

Project............ Date...............	Budget	Actual to date	Variance +/(-)	Est. Total	Comments
Phase 1					Phase to finish Jan 12
Project Management					
Consultancy					
Development					
Analysis					
Testing					
Implementation					
Site Preparation					
Hardware Configuration					
Administration					
Travel					
SUB TOTAL					
Project Overheads					15 % uplift
Contingency					10 % uplift
TOTAL					

Comments:

Signed:..**Project Manager**

Steps to planning a Resource Budget

Resource budgets are somewhat easier to ascertain, once the project plan and manpower budget has been identified. The only problem being that of how long will a team member need to utilize a resource? This can only be estimated by experience and common sense, given the skills and efficiency of the team member. The following is the proposed method:
- Ascertain resources, charge out rate and their available time frame
- Define Time and Work Allocation schedule *
- Allocate resources to each work allocation *
- Calculate the total cost for the resource by phase and total project. *
- Adjust team allocations and task duration until resource budget and project plan are both realistic, feasible and efficient. *

This is usually done via the project planning software

Planning Documents

Planning documents are some of the most important in your project.

Project Plan

Guide lines for usage:
Produced by Project Manager, based on analysis with client. It will be in text report format backed and supported by graphical project plans. It should include the updated Project Plan from the Project Review, the Staffing Plan, and Change Control issues and a Budget Review. It should also include updates on the following:
- Milestones passed
- Deliverables made
- Costs update
- Staff changes
- Changes to plan

Divide project plan into appropriate sections for each member of the project team, viz.:
- Individual's tasks and responsibilities
- Deliverables required
- Work schedule

Recipients:
- Project Manager
- Project Control Office
- Quality Manager
- Present appropriate parts of the Project Plan to project team for actioning
- Present Project Plan to client for information.

Report Contents
- Project Name - Include project mnemonic
- Project Sponsor - Identify the client department and the prime contact points within the department, stating their involvement with the project.
- Business Requirement - An overview only is required here.
- Major Deliverables and Milestones - List all deliverables and milestones with appropriate dates

- Project Stages - List all the stages and outline the time frame.
- Any major dependencies or constraints - highlight any significant or crucial factors that may affect the integrity of the project plan
- Project Plans - always a Gantt chart, showing tasks completed, to be completed and overdue tasks.
- Often a Pert chart showing inter-dependencies.
- Individual work assignments for project staff.
- Project Reviews - Up-to-date Project Reviews from project team, or a consolidation of same.

Staffing Plan

Guide lines for usage:
Produced by the Project Manager as part of the planning phase and to ensure efficient team management and control.

Recipients:
- Project Manager
- Project Control Office
- Quality Manager
- Sections to Client and/or Project Sponsor as appropriate

Report Contents:
- Staff Tree
- Staff Terms of Reference
- Numbers and time allocation of staff
- Costs of staff by staff type and period
- Proposed staff budget
- Details of any staff proposed.
- User staff involvement required.
- Updates
- Authorization

Work Allocation

Guide lines for usage:
A document produced by many planning software packages, or independently, that outlines each team member's tasks to be undertaken and the time frame assigned to this work.

Recipients
- Project Manager
- Project Control Office
- Quality Manager
- Appropriate team member

Report Contents
- Project
- Project Phase
- Staff Name
- Staff Group
- List of tasks and time frames
- Comments
- Authorization

Budget

Guide lines for usage
To plan, manage, control and report upon manpower and resources spend and proposed spend, with reasons for variances.

Recipients
- Project Manager
- Project Control Office
- Quality Manager
- Client and/or Project Sponsor

Report Contents
- Budget with respect to period, group and phase
- Variances
- Reasons for Variances
- Estimated future spend
- Authorization

Quality Plan
Guide lines for usage
The Quality Plan is the agreed actions, and plans to be instigated in order to meet the Quality Strategy and ensure that the products:
- meet the user's requirements
- meet agreed standards
- are complete
- Have all errors detected and corrected as soon as possible.

Once the Quality Strategy and Quality Review procedures have been established, the resources, time, cost and effort to undertake the work must be factored into the project plan and resource plan.

Recipients
- Steering Board
- Users
- Quality Assurance
- Project Manager
- Project Control Office
- Project Team - Design, Development, Testing etc.

Report Contents
- Project Name
- Project Manager
- Quality Manager
- Quality Strategy
- Quality Review methodology
- Project Standards
  - analysis and design methodology
  - documentation
  - coding
  - testing
- Product Descriptions (deliverables)
  - system
  - programs
  - hardware
  - operating software
  - documents
  - test specifications

- Schedule and independent audits
- Quality Resourcing Plan
- Quality Budget
- Testing procedures
- Acceptance Test Plan.

Test Plan
Guide lines for usage
As a planning document setting out the requirements, standards and time frames of the testing procedures

Recipients
- Quality Assurance
- Project Manager
- Project Control Office
- Design
- Development
- Testing
- Users - for User Acceptance Testing

Report Contents
- Objectives and standards to be met
- Include Quality standards, programming and documentation standards
- Types of tests
    - Program Unit testing
    - System testing
    - User Acceptance testing
- Describe the test scope and how to perform them
- Source of test data
    - Describe where the test pack will originate from and details of any parallel running to be undertaken.
- Responsibilities
    - Note names of the testers, users and managers and their respective responsibilities

- Testing Plan
  - A schedule of the expected effort to conduct each test module, the order of testing, and the testing assignments will be included in the test plan.
- Control procedures
  - Include details of error trapping, error reporting and change control methodology
- Quality Manger's Authorization
- Project Manager's Authorization

Controls Within a Project

What are They?

Requirements:
- project requirements
- user requirements
- system specifications
- functional specifications
- program specifications

Project Planning:
- quality
- risk
- resources
- budgets
- time
- deliverables

Implementation:
- change
- versions (hardware, software and configuration)
- releases (hardware, software and documentation)
- testing
- errors

What Types of Changes are There?

Major changes are those that will make a significant impact upon resources, budget, time scales, project plan or existing development or hardware configuration. Major changes will usually impact upon time scales, costs and sometimes contracts- for example a major database structure change.

Minor changes are those that have an insignificant impact upon the proposed system or project plan and will take minimal implementation - for example a help-screen change.

Planned changes are those that are anticipated within the project plan - for example a new version or release of software.

Unplanned changes are not within the existing project plan and usually originate from the user or from discovered error- for example a request for more system functionality or a correction of an error found in testing.

Anticipated changes may not necessarily be within the existing project plan, but they usually originate from consultancy and analysis discussions or from a proposal made by the Project Manager - fore example a hardware configuration change to take advantage of newly emerged technology.

Unanticipated changes are by far the worst kind, and usually signal poor project management or client relationship, both of which should highlight any possible requests or future requirements. Should an unanticipated change be requested, remedial action should be immediately taken to regain awareness of user's needs and control of the project. They can originate from user, project team, quality reviews or testing.

Controls in Action

Follow this route:
- Set up the baselines
- Measure the baseline - quantitatively or qualitatively
- Set the acceptability level
- Measure the variance
- Control the changes to the baselines
- Adjust plans to take into account the changes
- Report changes
- Document

Why Control Them?

A project is a living, changing, shifting, and evolving, complex environment.

During constant user interaction many changes to the planned system will evolve as the users become more aware of what is available to them.

In a long, complex project is difficult to fully define all the requirements at inception.

External factors such as technological advances or new business requirements may signal or necessitate changes to the technical platform, applications or usage of the proposed system

Effects of Lack of Controls

Poor planning - the largest reason for project failure is through lose of control of the direction of the project, its resource usage; risk and quality control which are all are interdependent

Confusion as to the direction and aims of the project. The Project Manager must be aware of changes to the existing project plan and specifications, to enable him to make the necessary decisions and adjustments. The project team must be aware of what their remit and terms of reference are in order to be able to function as a team. The users need to be aware and up to date to ensure their full and effective involvement.

Inefficiency with resource usage and the implementation of the system.
Divergence from the client's requirements and expectations

Ineffective Project Management and lose of command of the direction of the project, its resource usage, quality and risk

Poor communication - the users, project team and clients must be aware kept up to date with actions, deliverables and requirements of them, to enable the project to run smoothly.

Lack of confidence and disappointment in the project team. A badly controlled project quickly diverges from the requirements and expectations of the users and clients

Unnecessary difficulties are caused when the ramifications of a change have not been fully explored, documented or communicated. A change will have a ripple effect upon other tasks, actions or systems that all must be aware of.

Misdirection of resources and problems with the optimum planning of resource usage.

Extra costs and time allocation due to poor resource usage. To the detriment of User Acceptance and the credibility of the project and its sponsors.

Loss of functionality either through poor or ineffective testing and Quality Assurance or because lack of documentation means that software enhancements and upgrades are difficult. In addition, the changes, if not properly controlled, will be made without consideration of the impact or interaction upon other parts of the system.

Duplication of effort or of hardware, software and documentation configurations.

Summary

Lack of control causes:
- The largest reason for project failure
- Divergence from the client's requirements and expectations
- Loss of command of the direction of the project, it's resource usage, quality and risk undertaking
- Lack of awareness of what is happening
- Problems with the optimum planning of resource usage
- Loss of the functionality of the system
- Disappointment and lack of confidence in the project team.

Change Control

Overview
- Log change request
- Estimate
  - impact
  - risk
  - benefits
  - costs
- Plan the changes
- Make changes
- Quality Assurance
- Test
- Document
- User Acceptance
- Handover

When Change Control Should be Used

Change Requests can be instigated for minor changes or major changes.
- Major changes, those that will make a significant impact upon resources, budget, time scales, project plan or existing development or hardware configuration, must be authorised by the Steering Committee before actioning within the Change Control procedure. Major changes will usually impact upon time scales, costs and sometimes contracts.
- Minor changes, those that have an insignificant impact upon the proposed system or project plan and will take minimal implementation. may come from the Steering Committee or from an authorised user. It is unlikely that minor changes will impact upon costs or time scales.

Aims of the Change Control Procedure

The procedure is based on the Change Request Form (CRF), which acts as a means of communication, record keeping, and progress review.

It allows the user to:
- Be notified of the expected cost and the schedule of work.
- Plan for the implementation of the change.
- Reduce any disruption to a minimum.
- Plan any associated training, etc.
- Assure them that the implementation has been executed efficiently.

The Project Manager has a controlled means of changing the requirement that allows for better project planning and execution.

Whilst the specified goal may change, a baseline, against which the project success can be measured, is still in existence and known by all.

Actions

This procedure describes the way in which the Change Request From (CRF) is used as both record and communications medium.
- Origination - the User makes a formal request for a change by filling in page 2 of the CRF and obtaining the appropriate authorisation.
- Investigation - the nature and priority of the change is investigated by the Project Manager and authorised to proceed.
  - The front cover sheet is completed and a unique log number allocated to the change request.
  - After logging by the Project Control Office (PCO), the project plan, the resource allocation and budget is updated, as appropriate. Feedback from the responsible project leader may be required before these actions can be undertaken.
  - The work is then allocated to the correct project group, who become the responsible project group.
  - The project leader acknowledges the work allocation and thus accepts responsibility for its completion.

FastTrack © Project Management

- Authorisation - the project leader details the system, procedures and documents that will be involved within the change, and then authorises the changes to them.
- Estimation - the project leader then estimates the work to be undertaken and the work schedule
- Proposal - the project leader works with the Project Manger to produce a proposal that includes, estimated costs, impact and the impact of the changes (from the user's point of view).
    He signs the proposal, confirming the best endeavours of the project team and acceptance of undertaking the work within the proposed time and budget.

Authorisation - the appropriate user signs to authorise acceptance of the proposal.

Allocation - the PCO checks the authorisation, updates the project plans and then allocates the work to the appropriate responsible project group.

Acknowledgement - the project leader acknowledges the work allocation and accepts responsibility for its completion.

Work - the work is undertaken by the responsible project group, with appropriate updates to the project plan and project reports.

Completion - when the work is completed, the fact is confirmed by the signature of the project leader and responsible person.
Total costs and time allocations are noted, together with any updates on the impact of the change.

Quality Assurance - the changes are passed to the Quality Assurance group for their review and report. Corrections are made as required

Testing - the changes are passed to the testing group for final testing. Corrections are made as required.

Summarisation - the changes made are summarised, together with any deviation made from the proposal, by the Project Manager, who signs off the work as being ready for implementation and User Acceptance Testing.

User Acceptance Testing - the users undertake their User Acceptance Testing,

User Acceptance - when the users are satisfied with the changes made, then the appropriate user, signs to authorise User Acceptance and movement of the change into the live environment.

Confirmation - the Project Manager confirms that the implementation of the change into the live environment has taken place.

Handover - the user and, if appropriate, the support function, sign to agree to the acceptance of the change as their responsibility within the live environment.

Quality Control

What is Quality?

Quality....British Standard 4778 quotes "...the totality of features and characteristics of a product or service which bears on its ability to satisfy a given need." In the case of a project the "given need" is defined by the user's requirements.

A quality system, an often forgotten, but very important deliverable of a project, is:
- Error Free
- Well structured and defined
- Complete
- Comprehensive
- Flexible
- Efficient and Reliable
- Easy to maintain
- Easy to enhance
- Well documented

A good quality system increases client and user satisfaction and the lifetime of the system with a corresponding decrease in project and maintenance costs.

Quality Control

Quality Control is the examination and control, to an agreed standard, of the products of a project.

Quality Control can be divided into the following actions:
- Agreeing the quality criteria
- Planning and Resourcing quality reviews and testing
- Demonstrating the meeting of the quality criteria
- Undertaking changes in a controlled and documented fashion (Change Control)

Quality Control needs:
- Organisation
- Specific checking for quality
- Support from the Project Board or Steering Committee and the project sponsors
- Support from the user, client and senior management
- Understanding by user, client and project team
- To become an integral part of every project activity and task

Quality Strategy

Quality Strategy is an agreed documented statement of the methods and procedures, to be used within the life cycle of the project, in order to ensure that a quality product and system is produced

It should include standards for:
- Applications.
- Software configuration.
- Hardware configuration.
- Testing.
- Documentation.
- Training.

Quality Control can be divided into the following actions:
- Quality Strategy
- Quality Planning
- Methodology
- Resourcing
- Product Descriptions
- Quality Reviews
- Change Control

Quality Plan

The Quality Plan is the agreed actions, and plans to be instigated in order to meet the Quality Strategy and ensure that the products:
- Meet the user's requirements.
- Meet agreed standards.
- Are complete .
- Have all errors detected and corrected as soon as possible.

It includes:
- Quality Strategy
- Quality Review methodology
- Resourcing plan
- Budget
- Product Descriptions

Once the Quality Strategy and Quality Review procedures have been established, the resources, time, cost and effort to undertake the work must be factored into the project plan and resource plan.

Product Descriptions

What

A description of each identified product, deliverable or item to be produced by the project. It also details the standards to be met by the product in quantitative and qualitative terms.

Aim
- To provide a description of what is required
- To furnish a baseline to measure the finished product
- As an aid to planning the resources to produce the product.
- To provide a checklist to be used at quality review.

Action
Produce a description that includes the following agreed information:
- The purpose.
- The products from which it is derived.
- The product composition.
- Benchmarks and / or standards.
- The quality criteria.
- The quality control method to be used.

Quality File

The Quality File contains all the forms produced from the quality reviews during the life of the project. It ensures that a clear audit trail of work is in evidence. The independent Quality Assurance reviewers would investigate the file to ensure that the quality checks have been carried out correctly and competently. It should include the Quality Log - a unique numbering system for the paperwork.

Quality Reviews

What
Quality Review is the process by which a product, group of products or part of a product is checked against an agreed set of quality criteria.

Formal Reviews - A group inspection under very tightly defined procedures.
- Prior to the implementation of a major deliverable.
- at the end of each stage of the project
- At the end of the project, prior to User Acceptance.

The reviewers consist of representatives from users, the appropriate members of the project team, the quality assurance group and members from senior business management - to ensure adherence to business objectives.

Informal Reviews - A walk through - a "walk" by the product author of each individual step within the product. Or an inspection of a smaller, less strategic product. The checks and reviews applied concentrate on the following aspects:
- Technical correctness- does the product adhere to the agreed methodology?
- Business correctness - does the product reflect the user's perception of what it represents and what he wants?
- Fitness for purpose - does the product meet its objectives?

Aim
- To find any errors at the earliest time so minimising costs, disruptions and the escalation of errors
- Ensure that errors are corrected in a structured and documented way
- Remove omissions, ambiguities and errors in the system design and user specification
- By using a continuous staged monitoring process, to minimise time, disruption and effort.

Action
- Preparation - the administration of identifying the attendees, distribution of the product and its quality criteria and its study.
- Review - a formal meeting, to discuss errors found, their resolution and the plans and resources to allocate to the actions.
- Follow up -the correction of the actions, by Change Control, the confirmation that errors have been corrected, and that compensating errors have not been made. The formal sign off.

Risk Control

Areas of Risk

- Project Success. The overall success of the project is subject to the myriad of problems associated with such a complex environment. The problems can originate from the project team, the planning and management of the project or just insufficient resources. Excellent project management skills are the way to minimise risk.
- System Environment A system that may look good from the outside or from cursory glance, make be subjecting the clients to considerable risk by its design. The most common risk areas are that of loss of data from unlawful intrusion, a disaster or component failure. Risk can be minimised by the judicious use of security features.
- Implementation take-up. The best designed and controlled system will not be used properly after implementation if it is not what the client required or they are not equipped to make the best use of its facilities. Project Management and Consultancy skills to ascertain and update the client's requirements will minimise risk.
- External factors. The best planned, controlled and designed project may still fail due to external factors that range from supplier collapse to major reorganisations of your client's company. Knowledge of and immediate action to minimise the risk are the best weapons.

Risk Assessment

Not only is risk difficult to identify but the level of risk can only depend upon the identified threats and the assessment of the vulnerability the factor has to these threats.

There are many different ways of measuring risks, mostly statistically based. The best assessment of risk is avoidance by careful planning and controls and the knowledge of where potential areas are.

How to Minimise Risk

Project Success
- Control of budgets, time and resources.
- Build a good foundation for the project.
- Good project planning and design.
- Good, efficient, integrated team.
- Motivated project team and users.
- Constant reviews.

System Environment
- Appropriate to the user's needs
- Appropriate to the business and physical environment
- Secure
- Technically sound
- Upgradeable and Maintainable

Implementation take-up
- What the want
- Good support
- Efficient and effective technological transfer
- Generate enthusiasm

External Factors, avoidance of, or forewarning of:
- Force Majeure
- Corruption and Collusion
- Poor Supplier Choice
- Poor Support and Maintenance Company
- Project Sponsor failure
- Internal company restructuring

Requirements to Ensure Minimum Risk
- The agreement of the user 's needs
- Good Project Planning
- Effective Project Control
- Appropriate System Design
- Up to date Skills
- To know what is going on within the project environment

To Minimise Risk.....

Planning
Controls
Communication
Documentation
Experience
Awareness
Knowledge

Security

System Environment

Access to the system
- Locks
- Security or electronic surveillance
- Entry/Exit logging

Physical environment
- Fire fighting, prevention and exits
- Health and Safety measures
- Air Conditioning
- Electrics resilience and UPS
- Light duplication
- Heat measurement and monitoring
- Data storage
- Off Site storage
- Data Library access logging
- Consumables logging
- Site security
- Duplicate sites
- Appropriate environment
    - Force Majeure
    - Disaster Recovery

Hardware

Access to the hardware
- Encryption
- Power on password
- Locks and cables
- Logon traces and audit trails to the server, peripherals, network, terminal
- Access logging

Access to the peripherals
- Locks and cables
- Removal
- Access logging

Ruggedness of the hardware and comms
- Appropriate to the use and environment
- Expandable
- Non removable

Data storage
- Disc drive and tape locks
- Removal of storage device

Resilience of the hardware
- Sharing and/or exchanging of hardware
- Mirroring
- Shadowing
- Excess Capabilities
- Routine Maintenance
- Support Software for optimum configuration

Resilience of the comms
- Good design
- Duplication of comms
- Modular design

Force Majeure
- Disaster Recovery

Software

Access to the software
- Access logging
- Access denial
- Audit trails
- Power up passwords
- Logon timings
- Logon restrictions

Access to the peripherals
- Access logging
- Access denial
- Audit trails
- Power up passwords
- Logon timings
- Logon restrictions
- Consumables control

Ruggedness of the software
- Housekeeping routines
- Maintenance routines
- Duplication of software
- Encryption
- Change control
- Version control
- Error control
- Release control
- Documentation

Data control
- Grandfather - Father - Son back-ups
- Interim Back-ups
- Full Back-ups
- Archiving
- Software library access logging
- Off site storage
- Duplication of back-ups

Force Majeure, Disaster Recovery

Staff

Collusion and Corruption
- Pre hire checks and references
- Frequent staff rotation
- Adequate compensation
- Staff control and management

Reliance on key Personnel
- Duplication of duties
- Frequent staff rotation

Abilities
- Appropriate hiring
- Frequent training
- Incentives
- Staff monitoring

Force Majeure, Disaster Recovery

Implementation Controls

Implementation, after Project Inception is probably the most stressful and difficult of the project stages, you and your team are tired, much of the excitement of the project has disappeared and people are looking to the next project.

Your clients and users will also find the implementation stressful and difficult. The following are the usual reactions to implementation:

- Impatience - you have to resist all the pressures of "going live" until the system and the preparations are totally ready.
- Apprehension - fear of the unknown or fear of the perceived known. These people must be motivated, reassured and pointed in the right direction.
- Perfectionism - you have to decide if the continual testing, checking and communication from your users is a trait of perfectionism, a message that there are problems with you testing and QA controls or a sign of apprehension.
- Unwillingness to Commit - There is a group of people that are totally unable to make a commitment to any change or decision. Hopefully by this stage of the project these traits have been identified in your users. Motivation, assurance, training and technology transfer is the answer.
- Unwillingness to accept responsibility - again an unfortunate personal trait. Or is it that the system does not meet the user's requirements. Careful listening and analysis to the reasons and rationale given must be undertaken.
- Being unprepared - is it poor project management, inadequate training, lack of interest or avoidance of the issues. Again careful listening and analysis is needed.
- Suspicion- of what the implementation will involve - either it's too good to be true or conversely they are worried that there is a secret agenda - maybe their job or responsibilities will adversely change. Communication and reassurance are the weapons here.

- Conflict - can arise from fear of the unknown or change. It could also be a signal that the user is not getting what he wanted or expected. In any project implementation there is always organisational and structural changes within the user business environment. The Project Manager must ensure that he overcomes these problems, whilst remaining independent and free from office "politics." A good Project Manager must use all his skills and personality to overcome these problems.
- Lack of support - maybe it isn't the system the users wanted - they aren't ready or are more interested in their current work. Maybe their existing workload is too high. Communication and reassurance, coupled with working to rectify the problems are the weapons here.
- Lack of interest - People often find a way of overcoming the failings of a badly designed system, especially when it fails to adapt to required changes. It thus loses user support and approval, who then devise alternative and unofficial systems to cope with the new requirements. The installed system then becomes bypassed, despised and falls into disuse.

The Implementation Procedure

Planning - should be an integral part of the Project Plan
   - Updated for the current situation?
   - Potential problems built into the project plan
2. *Preparation* - for implementation
   - The impact and implications of the changes
   - Marketing and Promotion
   - Site preparation complete?
   - Training to a sufficient level?
   - Users aware and ready?
   - Project Team aware and ready?
   - User Acceptance?
3. *Staging* - of the various implementations
   - Technically correct order?
   - Maximise enthusiasm?
   - Maximise effectiveness and efficiency?
   - Minimise cost and time frame?
4. *Implementation*

5. *Evaluation and Adjustment* - of the implemented systems and their procedures
   - Review meetings
   - Quality Assurance
   - Implementation Reports
   - Consultancy and feedback
6. *Replanning* - to take into account what has been learnt
7. *Consolidation* - of the happenings, wider implications of the implementations and the feedback
8. Return to Preparation (2) until all stages are implemented and consolidated.

Change Management

Change Management (as opposed to Change Control) is the management of the movement from one technical environment to another and/or from one system to another.

Change Management happens during a period of major structural and organisational change, thus in an ever changing and shifting environment, fraught with confusion, requirements must be analysed and undertaken.

It is probably the single most difficult skill that a Project Manager must learn. It can only be achieved by:

- ☑ **Commitment**
- ☑ **Motivation**
- ☑ **Concern**
- ☑ **Planning**
- ☑ **Feedback**
- ☑ **Experience**

Change Management Problems

"We've always done it this way." No Project Manager has can class himself as experienced until he has heard this phrase at least one hundred times - which is approximately five project implementations.

The memory of past failures tends to be bought at this stage - usually to the detriment of the changes required in this project.

Lack of confidence in the system, themselves or the project team.
Incorrect or inappropriate expectations of the system functionality.

Fear or disapproval of how the user perceives his responsibility and authority will change.

Organisational changes inherent to system implementation make the identification of the appropriate user of expert difficult and complex.

Each person's acceptance of change varies with the following:
- The impact of the change upon them
- Their own enthusiasm and that of their peer group
- Their person knowledge of the change and their proposed job
- The gain or loss of their personal esteem and professional value being bought about by the change.
- Their perception of the change
- Their technical skills
- Their age and socioeconomic group

Post Implementation

At post implementation the following should occur:
- Formal wind up of the project
- Completion of all documentation
- Staff appraisals
- Collation and analysis of all the existing data for use in future projects
- Post Project Review with the clients to ascertain any potential problems
- Formal release of project staff
- Thank you's all round
- Payment for final deliverables
- Possible sales pitch for future work.

Version Control

Overview

Version - A release of a software program or hardware configuration that has its own unique functions and specification. Usually each succeeding version employs enhancements and improvements to the prior. Version X.y is the usual syntax used, where X signifies a release that includes enhancements or improvements. y signifies a minor version change to correct bugs and errors.

Version Control - Used to ensure that a software program or hardware configuration is formally documented, recognised and controlled and that more than one version at a time, of a program is not in existence in the live environment

Version control is undertaken by the programmer or system development team leader prior to the installation of the software in the live environment. Version Control can also be used within the development environment, if there is a significant amount of software being developed, or there is more than one programmer. Differentation is made between development versions and live versions.

The purpose of version control is to ensure that:
- new versions are differentiated from previous releases
- each version is installed in the correct order
- changes from one version to the next are documented
- the latest version is available within the live environment
- modules from different versions are not muddled with possible disastrous results

Release Numbering

To assist with the identifying the release identity of installed objects it is recommended that the following convention be used.

All objects within a release should be assigned a *major* and a *minor* release identity code.

The major code may be in the range 01 ⇨ 24. The minor code may be in the range 00 ⇨ 59.

The major code should only be incremented where the release itself includes new functionality. The minor code should be incremented for any interim releases, bug fixes or minor enhancements.

Once the release identity code has been established, each object of the release should have its' time stamp altered to reflect the release identity - i.e. every file will have a DOS time attribute which reflects the release level the file was introduced.

In addition each object of the release should have its' date stamp altered to reflect the release date -i.e. every file will have a DOS date attribute which reflects the date the file was introduced to the system.

For example - the Windows System version 3.10 release is controlled in this manner...

```
MORICONS        DLL     118864      20/03/02    3:10
PBRUSH          DLL       6766      20/03/02    3:10
RECORDER        DLL      10414      20/03/02    3:10
```

Deliverables Schedule

This document should form part of every release of software (no matter how minor).

The schedule will show precisely which objects within the release are new to the system and which objects are being updated.

Version Isolation

As part of the preparations for making a release of software, the developer must isolate and 'freeze' all associated source, object and data files.

The entire development libraries should be archived and stored both on and off-site. If space permits three versions of source should be kept on-line i.e. previous release, Current release and the ongoing development source.

The entire testing libraries should be archived and stored both on and off-site.

Two versions of the test-bed should be kept on-line i.e. Current live release and the ongoing development test-bed. In this way ongoing development can continue without impact on customer support and fault detection.

Release Control

The development team, in the case of a software module, or the hardware installation team, in the case of a new hardware configuration, or unit being installed into the live environment undertakes release control.

In the case of a large project team organisation the implementation team undertakes the control, with liaison with the development of hardware team.

The Project Control Office will monitor the documentation. If there is a configuration librarian, then this person will be charged with overall management of the control function.

The Project Manager has overall signing authority of the release of the new module or unit.

The aim of Release Control is to ensure:
- that the release of hardware and software elements to the live environment:
- It is planned and controlled.
- Release is made at the optimum time to ensure minimal delay, expense and disruption.
- Releases are only made after full User Acceptance.
- that releases includes the following documentation:
    - Comprehensive installation/upgrade procedures.
    - Instructions for data file or configuration conversions.

- Deliverables schedule identifying each object in the release.
- Full updates of User, Maintenance and Support documentation.
- A completed Release Control Report.
- Formal acceptance from the support function.
- Formal acceptance from the User.

Actions
- Confirm that User Acceptance has been formally given for the release
- Confirm that all documentation has been updated
- Plan and cost the optimum release method and date
- Update the project plan
- Complete the Release Control Report
- Release the unit or module
- Test in live environment
- Quality Assurance
- Formal Hand-over

Testing Control Procedure

Overview

This procedure covers the three different type of testing which are undertaken during the project life cycle - unit testing, system testing and user testing.

By following the procedure:
- The Project Manager gains a more controlled means of notification and correction of bugs and/or errors during development and implementation of the requirements.
- The Project Manager assures himself that the development is adhering to the Quality Assurance standards and the User Requirement Specification (URS).
- The users assure themselves that the system meets the needs specified in the User Requirements Specification (URS)
- The users assure themselves that the system is robust and of good quality.

Unit Testing

The purpose of unit testing is to qualify individual components of a system.

Unit testing is conducted by the programmer responsible for productions of the unit under test, generally the person who has programmed the unit.

Actions
- Ensure that the unit meets the program specification
- Ensure that the results of any calculation match those that are expected.
- Prove that unit does not fail if inappropriate actions are suffered, e.g. invalid data entered. The unit must at least respond with a meaningful message advising the user of the nature of any erroneous condition encountered, and if possible should offer or take corrective action.
- Document the testing performed.
- If the unit passes, the source code will be marked as "read only" to avoid accidental damage and the tester will advise the Project Manager that the unit is ready for system testing.
- The Project Manager will determine when that unit will be transferred to the test environment for future system testing.
- If the unit fails the tester will, if the tester is the programmer, determine the correction required and apply that correction.
- If the tester is not the programmer, the tester will tell the programmer that the unit has failed, and why. The programmer will then determine the correction required and apply the correction.
- In either case the Project Manager will be advised of the failure by the tester and of the expected time to fix by the programmer, and an expected date of next testing.
- Copies of the test reports will be filed with the program specification and sent to the Project Manager for review.

System Testing

The tester may be anyone who has not programmed the part of the system under test. If more than one tester is available, each may test the other's work; otherwise the Project Manager will appoint another person.

The purpose of system testing is to qualify the system as a whole, by testing the menu structures, user and program interfaces, screens, software modules, etc.

The procedure for the system testing is based on the Test Plan, which is prepared in the Design phase of the project life cycle.
- Review the method and expected results described in the plan.
- Follow the test method, recording progress through each test and the test results
- The expected and actual results are compared by the tester and if within limits, that system part will be deemed to have passed.
- The system passes only if all parts pass.
- Document the testing performed.
- Pass the test reports to the Project Manager for review.
- If the system passes, the Project Manager will advise the user's contact that the system is ready for user acceptance testing.
- The programs and any control files will then be version stamped and transferred to the user test environment.
- If the system fails the testers will report the failure to the Project Manager who will discuss the issues arising with the development team. Together, the Project Manager and programmers will determine what corrective action will be necessary and when the system will be next ready for system testing.

User Acceptance Testing

The user department will be responsible for providing suitable test data or *shadowed* live data for the purpose of testing.

The purpose of user acceptance testing is to qualify the system as a whole, to the satisfaction of the Project Sponsor, ensuring that the Project Goals have been met and the User Requirements satisfied.

The procedure for user acceptance testing is also based on the Test Plan, which is prepared in the Design phase of the project life cycle.
Actions:
- A period of parallel running of the old and new system is usually undertaken
- All modules and units of the system will be tested and compared to the System Specifications.
- Where anomalies are encountered the tester will initially discuss the issue with project team representative and if a resolution is not forthcoming raise an Error Report documenting the problem.
- Upon completion of the testing all Error Reports will be submitted to the project team for review.
- The Error Report will be reviewed by the project team
- An Acceptance Test Review meeting will be held with the development team, support and the User department. At this review any outstanding issues can be discussed.
- If the Project Sponsor is satisfied with the system, then User Acceptance will be instigated.

Error Control

An internal control mechanism and means of communication, used to capture, notify, resolve and document errors found during quality review or unit, system or user acceptance testing.

Project team members involved within quality review or unit, system or user acceptance testing undertakes error control.

Error Control aims to be the control mechanism to ensure that:
- All errors are captured and resolved.
- Their resolution does not incur compensating errors.
- All errors are addressed in the most logical and cost effective manner
- Progress, costs and resources are not unduly compromised.

It should be an internal reporting mechanism for:
- The formal capture and documentation of all errors originating from quality control or testing. Regrettably errors within the live environment may also be found.
- The provision and communication of the complete specification of the errors together with their impact on all modules, units and interfaces.

Actions:
Identify error
- Allocate a unique error log number
- Fully document error and it's impact upon other development work, software modules and the project plan
- Estimate the costs and the required resources
- Obtain Project Manager authorization
- Update project plans
- Undertake error correction
- Test
- Quality Control
- User Acceptance
- Sign Off

Implementation Control Reports

Change Control (CRF)

Guidelines for usage
A formal method of control used whenever any change is requested to hardware, software, documentation, and contract of project plan.

Activities surrounding its usage:
- Log change request
- Estimate
  - Impact.
  - Risk.
  - Benefits.
  - Costs.
- Plan the changes
- Make changes
- Quality Assurance
- Test
- Document
- User Acceptance

Recipients
- User, Steering Board, Client
- Project Manager
- Project Control Office
- Development
- Quality Assurance
- Testing
- Project Team as required

Report Contents
- Project Phase or System
- Date
- Log Number
- Instigator
- Priority
- Description of required change
- User Authorization
- Change Definition
- Project Manger Authorization
- Project Control Office Check
- Allocation
- Project Plan for the change

FastTrack © Project Management

- Estimated costs
- Impact of the change
- Project Manger Authorization
- User authorization
- Project Control Office Check
- Work Allocation
- Allocation acknowledgment
- Work completed
- Cost , time and impact notification
- Quality Assurance
- Testing
- Project Manger Authorization for implementation
- User authorization
- Project Control Office Check
- Installation Confirmation
- User Acceptance

Error Report

Guidelines for usage
An internal report for the correction of errors originating from, quality control or testing.

Recipients
- Project Manager
- Project Control Office
- Development
- Quality Assurance
- Testing
- Project Team as required

Report Contents
- Project Phase and Task
- Date and Log Number
- Priority
- Instigator
- Error
- Affected modules, units and interfaces
- Status
- Requirements
- Estimated Costs
- Completion sign off
- User Acceptance

Version Control Report

Guidelines for usage
The purpose of version control report is to:
- Ensure that software versions are differentiated.
- Installed in the correct order.
- Have the changes from one version to the next documented.
- Ensure that the latest version is available within the live environment.
- Ensure that modules from different software versions are not muddled with possible disastrous results.

Differentation is made between development versions and live versions.

Recipients
- Programmer
- Project Manager
- Project Control Office
- Quality Assurance
- Support function

Report Contents
- Version Control Procedures
- Release numbering
- Deliverables schedule
- Version isolation procedures

Release Control Report

Guidelines for usage
Whenever a new software module is released into the live environment.

Whenever a new hardware module or hardware configuration is released into the live environment.

Recipients
- Project Manager
- Project Control Office
- Quality Assurance
- User Support section

FastTrack © Project Management

Report Contents
- List of Contents
- Version Identification
- Description of new and changed functions
- List of Error Reports fixed in the release
- Technical Architecture
- System overview
- Deliverables schedule identifying each object in the release.
- Installation or Upgrade Guide
- Instructions for data file or configuration conversions.
- Training material updates
- User, Support and Operations Manual updates
- Help desk or support function details
- Formal acceptance from the support function
- User Acceptance sign off

User Acceptance

Overview

User Acceptance - The formal agreement to accept an implemented system into the business environment. It includes agreement that the system has met the User Requirement Criteria. Responsibility for the system passes from the project team to the client.

User Acceptance Sign Off - The formal documented signature accepting responsibility for a system and signalling satisfaction with the project deliverables.

User Authorisation - the formal authorisation that either accepts a proposal or a change or allows the movement to the next stage or task within a project. It is an important control on User's Requirements. Often used as a checkpoint of understanding.

Why is it so Important?

Without User Acceptance:
- The system cannot be implemented
- You cannot be sure that User Requirements have been met
- The consultancy team cannot be paid
- The project or phase cannot be completed

The Procedure

Who
- The authorised user
- The Project Sponsor
- The Project Manager
- The appropriate members of the project team

Aim
- To ensure that the system meets the User Requirement Specification
- To pass responsibility for the system to the Users
- To permit the project to be wound up

FastTrack © Project Management

Actions
- After User Acceptance testing the sponsor will be invited to sign-off the project.

> **The *sign-off* will constitute the formal hand-over of the system to the User department.**

- The system support function will be asked to formal accept the support responsibility for the system.
- Where the Project Sponsor is not prepared to accept the system as a result of any identified anomalies, the project team will process the Error Reports or Change Requests raised, subject the system to unit and system testing and Quality Assurance and then re-submit the system for User Acceptance Testing.

User Acceptance Sign Off Report

Guidelines for usage
As a formal User Acceptance Sign Off and project closure action.

Recipients
- The Project Sponsors
- The Project Manager
- The Quality Assurance Manager
- File Copy

Report Contents:
- Project Name
- Executive Summary
- Overview of Aims of project
- Review of costs and time scale of project against estimated and projected figures
- Overview of problems encountered
- Overview of changes made
- Any future proposals
- Formal sign off of Project Manager
- Ramifications of formal acceptance
- Formal User Acceptance
- Invoice

Requirement Control Reports

Project Initiation Document (PID)

Guide lines for usage

It describes:
- What the project aims to do and how
- What the system aims to do and how
- Existing data usage
- Proposed data usage

Produced by Project Manager with input from consultants and analysts - based on further discussions with management/client. Is more detailed than the Project Charter and includes an executive summary and proposed outline schedule. Feeding from the Project Charter, Terms of Reference and Statement of Scope and Objectives it is written as a detailed project overview.

Sets up the initial planning criteria for the information of client, user and project team, it becomes the base-line for project progress control purposes.

Unlike the Project Plan it is **not** a living document and remains unchanged throughout the life of the project.

☑ **Establishes the baseline starting position of the project.**

Recipients
From Project Manager
- To Client
- To Steering Committee, or similar sanctioning board
- To the Project Team after client's signature

Report Contents
Project sponsor
- Identify the client department and the prime contact points within the department, stating their involvement with the project.

Executive summary
- This section should provide in broad terms a description of the project, identifying the client and describing the client's current problems. Where applicable, the summary should indicate any proposed approach to be taken by the consultancy, together with an indication of the priority which has been assigned to the project.

Business Requirement
- The business need for the project should be identified and documented here.

Terms of Reference
- The responsibilities of the clients, users and project team.

Project Scope and Objectives
- The definition of the objectives of the project, its background, reasons for being and constraints on the solution. The limits of the project as currently perceived. In addition to identifying which aspects of the project will be referenced the scope should also highlight any aspects which will not be considered.
- Where applicable, project phases and delivery tranches should be identified and scheduled here. Any time scales given are to be quoted ± 50% accurate.

Business Case
- Detail how the client and their business will benefit from the project implementation

Ramifications of Project
- Details of how adoption and implementation of the project will affect the sponsoring department environment and other departments.

Feasibility
- Findings from the SWOT analysis

Risk and Cost analysis
- Review of Possible Solutions (at least 4 including "no action")
- Costs and Benefits of each Solution
- Recommended Solution
- An outline of the risks and costs involved with the project and proposals as to how they can be minimized.

Project Organization and Responsibilities
- An organization chart and outline of the roles of each

Proposed schedule
- A schedule for the remainder of the project should be provided

User Requirements Specification
Guide lines for usage
The fully documented and agreed User Requirements to meet the following:
- The Business Requirement
- The proposed budget and resources
- The proposed costs of the system and it's operating costs
- The required system and technical architecture appropriate to present technology
- The system:
    - To meet the User's expectations and requirements.
    - To save the clients money and make them more profitable.
    - To improve productivity and efficiency.
    - To provide better management information allowing for better decisions to be taken.

Unlike the Project Plan it is **not** a living document and remains unchanged throughout the life of the project.

☑ **Establishes the baseline of the Project Requirements**
All other changes are made through Change Control

FastTrack © Project Management

Recipients
- Project Manager
- Steering Committee
- Project Sponsor
- Senior Users
- File Copy

Report Contents
- Executive Summary
- Computer Strategy
- Business Constraints and Dependencies
- System Objectives
- Existing Workflow. This section should document the existing work flow and operating procedures related to the project. Where applicable this text may be supported by use of suitable diagrams.
- Proposed Workflow. This section documents the proposed work flow and operating procedures, highlighting the processes that will require further development. Where applicable this text may be supported by use of suitable diagrams (DFD, ERD etc.)
- Specific System Requirements
    - Technical Platform
    - Hardware
    - Software
    - Functionality
    - look and feel
    - Structure
    - Interfaces
    - Dependencies
- Comms
- Maintenance and Support
- Housekeeping and Audit
- Security
- Environment
- Data transfer
- Required Benchmarks. This section documents all screen and report layouts for the system. Where applicable reference may be made to any prototype that has been built.
- People Based Requirements
- Procedures and Methods
- Users Involvement

- Training and Technology Transfer
- Organizational Based
  - Structural and Organizational
  - Jobs
- Summary of Problems
- User Authorization

System Specification
Guide lines for usage
As an agreed formal documentation of the system description.

Recipients
- Project Manager
- Steering Committee
- Development
- File Copy

Report Contents
- System Overview. The overview section describes the functionality of the system.
- Technical Architecture. The hardware, operating system and software environment are described in this section.
- Data Flow. Data flow diagrams and/or entity relation diagrams should be included in this section to describe the logical relationship of the data entities to be accessed in the system.
- Database Structure. The physical structure of the system data files should be documented here.
- System Hierarchy. A top-down analysis of the system showing all procedures and functions used in the system should be included here.
- System Inter-dependencies. Any inter-dependencies between this system and other systems must be documented here. These may include data exchange, system interfaces and data sharing etc.
- System Interfaces
- Benchmarks. Details of any benchmarks by which the system can be measured, whether provided by the client or as appropriate to the implemented system.
- Authorization

Functional Specification

Guide lines for usage
An agreed technical specification and description of the proposed system, outlining its function, produced for the benefit of the systems analyst.

Recipients
- Project Manager
- Senior User
- Development
- File copy

Report Contents
- Executive Overview
- System Overview
    - Description
    - Data flow diagrams
    - Inputs
    - Outputs
    - Interfaces
    - Security and Back-ups
    - Special hardware or software requirements
- Development and Implementation Plan

Program Specification

Guide lines for usage
An agreed technical specification and description of the separate programs within the system, produced for the benefit of the systems analyst.

Recipients
- Project Manager
- Senior User
- Development
- File copy

Report Contents
- Executive Summary
- Overview. Provide an overview of the functionality in the program, indicating the functional role of the program within the system.

- Program "called by". Where applicable provide a list of all other programs and modules which invoke the current program.
- Program "calls". Where applicable provide a list of all other programs and modules which are invoked by the current program.
- Files used. Where applicable provide a list of all data files accessed by the current program, indicating against each the type of access required.
  Functional description. Provide a detailed and comprehensive description of all functions and procedures within the program.

Requirement Control Reports

Project Initiation Document (PID)

Guide lines for usage
It describes:
- What the project aims to do and how
- What the system aims to do and how
- Existing data usage
- Proposed data usage

Produced by Project Manager with input from consultants and analysts - based on further discussions with management/client. Is more detailed than the Project Charter and includes an executive summary and proposed outline schedule. Feeding from the Project Charter, Terms of Reference and Statement of Scope and Objectives it is written as a detailed project overview.

Sets up the initial planning criteria for the information of client, user and project team, it becomes the base-line for project progress control purposes.

Unlike the Project Plan it is *not* a living document and remains unchanged throughout the life of the project.

☑ **Establishes the baseline starting position of the project.**

FastTrack © Project Management

Recipients
From Project Manager
- To Client
- To Steering Committee, or similar sanctioning board
- To the Project Team after client's signature

Report Contents
Project sponsor
- Identify the client department and the prime contact points within the department, stating their involvement with the project.

Executive summary
- This section should provide in broad terms a description of the project, identifying the client and describing the client's current problems. Where applicable, the summary should indicate any proposed approach to be taken by the consultancy, together with an indication of the priority which has been assigned to the project.

Business Requirement
- The business need for the project should be identified and documented here.

Terms of Reference
- The responsibilities of the clients, users and project team.

Project Scope and Objectives
- The definition of the objectives of the project, its background, reasons for being and constraints on the solution. The limits of the project as currently perceived. In addition to identifying which aspects of the project will be referenced the scope should also highlight any aspects which will not be considered.
- Where applicable, project phases and delivery tranches should be identified and scheduled here. Any time scales given are to be quoted ± 50% accurate.

Business Case
- Detail how the client and their business will benefit from the project implementation

Ramifications of Project
- Details of how adoption and implementation of the project will affect the sponsoring department environment and other departments.

Feasibility
- Findings from the SWOT analysis

Risk and Cost analysis
- Review of Possible Solutions (at least 4 including "no action")
- Costs and Benefits of each Solution
- Recommended Solution
- An outline of the risks and costs involved with the project and proposals as to how they can be minimized.

Project Organization and Responsibilities
- An organization chart and outline of the roles of each

Proposed schedule
- A schedule for the remainder of the project should be provided

User Requirements Specification
Guide lines for usage
The fully documented and agreed User Requirements to meet the following:
- The Business Requirement
- The proposed budget and resources
- The proposed costs of the system and it's operating costs
- The required system and technical architecture appropriate to present technology
- The system:
  - to meet the User's expectations and requirements
  - to save the clients money and make them more profitable
  - to improve productivity and efficiency
  - to provide better management information allowing for better decisions to be taken

FastTrack © Project Management

Unlike the Project Plan it is **not** a living document and remains unchanged throughout the life of the project.

☑ **Establishes the baseline of the Project Requirements**
All other changes are made through Change Control

Recipients
- Project Manager
- Steering Committee
- Project Sponsor
- Senior Users
- File Copy

Report Contents
- Executive Summary
- Computer Strategy
- Business Constraints and Dependencies
- System Objectives
- Existing Workflow. This section should document the existing work flow and operating procedures related to the project. Where applicable this text may be supported by use of suitable diagrams.
- Proposed Workflow. This section documents the proposed work flow and operating procedures, highlighting the processes that will require further development. Where applicable this text may be supported by use of suitable diagrams (DFD, ERD etc.)
- Specific System Requirements
    - Technical Platform
    - Hardware
    - Software
    - Functionality
    - look and feel
    - Structure
    - Interfaces
    - Dependencies
- Comms
- Maintenance and Support
- Housekeeping and Audit
- Security
- Environment
- Data transfer

- Required Benchmarks. This section documents all screen and report layouts for the system. Where applicable reference may be made to any prototype that has been built.
- People Based Requirements
- Procedures and Methods
- Users Involvement
- Training and Technology Transfer
- Organizational Based
  - Structural and Organizational
  - Jobs
- Summary of Problems
- User Authorization

System Specification

Guide lines for usage
As an agreed formal documentation of the system description.

Recipients
- Project Manager
- Steering Committee
- Development
- File Copy

Report Contents
- System Overview. The overview section describes the functionality of the system.
- Technical Architecture. The hardware, operating system and software environment are described in this section.
- Data Flow. Data flow diagrams and/or entity relation diagrams should be included in this section to describe the logical relationship of the data entities to be accessed in the system.
- Database Structure. The physical structure of the system data files should be documented here.
- System Hierarchy. A top-down analysis of the system showing all procedures and functions used in the system should be included here.

- System Inter-dependencies. Any inter-dependencies between this system and other systems must be documented here. These may include data exchange, system interfaces and data sharing etc.
- System Interfaces
- Benchmarks. Details of any benchmarks by which the system can be measured, whether provided by the client or as appropriate to the implemented system.
- Authorization

Functional Specification
Guide lines for usage
An agreed technical specification and description of the proposed system, outlining its function, produced for the benefit of the systems analyst.

Recipients
- Project Manager
- Senior User
- Development
- File copy

Report Contents
- Executive Overview
- System Overview
    - Description
    - Data flow diagrams
    - Inputs
    - Outputs
    - Interfaces
    - Security and Back-ups
    - Special hardware or software requirements
- Development and Implementation Plan

Program Specification
Guide lines for usage
An agreed technical specification and description of the separate programs within the system, produced for the benefit of the systems analyst.

Recipients
- Project Manager
- Senior User
- Development
- File copy

Report Contents
- Executive Summary
- Overview. Provide an overview of the functionality in the program, indicating the functional role of the program within the system.
- Program "called by". Where applicable provide a list of all other programs and modules which invoke the current program.
- Program "calls". Where applicable provide a list of all other programs and modules which are invoked by the current program.
- Files used. Where applicable provide a list of all data files accessed by the current program, indicating against each the type of access required.
- Functional description. Provide a detailed and comprehensive description of all functions and procedures within the program.

Communication

Communication is the lifeblood of your project. Let's look at how to use it effectively.

With your Clients

Consultancy

Consultancy is a high level form of analysis that allows baseline project definitions to be obtained, whilst imparting specialist knowledge to the clients. The Project Manager also tends to learn from the clients.

Client Relationship

Client Relationship - The very important dialogue and correspondence with the project sponsor and senior users that ensures:
- That the project runs smoothly.
- That agreement on objectives should be achieved
- Availability of key staff and resources when required
- Assistance with Change Management - driving the requirements from the top
- Commitment to the project's aims.

Steering Committee

The Steering Committee, sometimes called the Project Board, is the ultimate project authority; it is made up of senior executives, the project sponsors (or their representative), the Quality Assurance Manager and senior representatives from the users and business environment.

It is responsible for:
- Guidance to the Project Manager.
- Ensuring that the company's business and IT strategies are followed.
- Ensuring that the Business Requirements of the project are met.
- Ensuring that the requirements and the progress of the project are disseminated at a high level within the company.

- Budget authorisation.
- Resource authorisation.
- Project initiation.
- Project review.
- Project monitoring.
- Formal project closure.

Project Progress Reports

The most efficient and satisfactory way of ensuring productive feedback to your clients is by regular, standardised, written and informative Project Progress Reports.

They should be pertinent and concise and have a standard format for readability. The reports may also be sent to the Quality Manager.

Not reporting progress will cause the following:
- Loss of confidence in the project team
- Problems with obtaining adequate and appropriate resources
- Problems with Change Management
- Lower levels of co-operation from the clients
- Clients following diverse paths of action
- Disorientation and isolation amongst the project team
- An incorrect perception of the Business Requirement

With your Project Staff

Project Reports

An internal project control conducted on a regular basis, usually weekly, produced by a team member for the benefit of the Project Manager. Where there is a large project team, an upward pyramid of reports can be used, via the project leaders.

It is a formal report used to collate information of progress, best estimates where a deviation has occurred, problems and issues outstanding. It includes progress against the project plan and the quality plan.

FastTrack © Project Management

It is used as a method of project team members to keep each other and the Project Manger informed of their past, current and future actions.

It always includes a report on work done and in progress (WIP) and often includes Time Sheets and individual (or project leader) project plan updates.

The report is used to update the Project Plan and is used as formal documentation on project progress. By its nature it is not used externally to the project.

External information, for example of products can also be disseminated. If an electronic bulletin board is available the review is often posted on it.

The contents of the report should support the team member's time sheets, last reports, project updates and time and work allocation update.

Honesty in the reports must be greatly encouraged.

Project Meetings

Progress reports serve as a one way only means of communication, a project meeting allows information to be disseminated and assimilated by all the team.

They must be regular and carefully planned with an appropriate agenda. The Project Reports and updated Project Plan should be the basis for discussions.

The meetings must be carefully controlled and documented to ensure efficiency of use of time. Friday afternoon meetings concentrate the mind on the tasks ahead, as everyone is keen to leave for the weekend.

Discussion from all should be encouraged, but in a controlled and effective manner so that "hobby horses", "soap boxes" and arguments about minutia are discouraged.

The proper involvement of all team members can lead to motivation, several views on a subject and the agreement of all to a course of action.

The meetings should also be used to the allocation of next week's work and tasks at hand as well as the investigation of work completed and any slippages.

Time and Work allocation

These are documented task lists, often produced by the project planning software, which are given to each team member, on a continuing basis throughout the project.

In a large project they would take the form of a list of macro tasks, supported by 1-2 weeks detailed tasks and time-scales.

They should be updated within the Project Reports and Meetings and form the basis of the Project Plan update

Staff Reviews and Meetings

Internal meetings with the Project Manager or project leaders should also be regularly held to discuss staff performance, training and guidance needed. Needless to say, praise and encouragement should also be included.

This should assist with staff motivation and training, but also identify any staffing problems.

With the Users

It is important to maintain regular formal and informal contact with your users.

Poor communication causes the following:
- A high degree of uncertainty amongst the users
- The users' fear they may lose control of their project requirements
- Problems with Change Management
- Lower levels of co-operation from the users
- Users following diverse paths of action
- The possibility of development of informal systems

- Disorientation and isolation amongst the project team
- An incorrect perception of the users' needs
- Loss of confidence in the project team

User Meetings

Regular formal or informal two-way meetings with the users and their management should be held to ensure that they all know:
- Project progress
- Updates on resource usage and budgets
- Requirements and deliverables needed from the users
- Feedback on user activities to assist with project planning
- Any change in requirements
- Any organisational changes

User Questionnaires

Questionnaires are a list of appropriate questions used within data gathering in the following cases:
- When a large statistical survey needs to be taken
- When job descriptions are required
- To obtain uniformity of approach
- To save time
- To obtain feedback from many users
- To ensure that the users feel involved

They may also be used in face to face analysis - either as a prompt or as a control on the uniformity of questions asked.

The problems of questionnaires:
- Production, distribution, collection, monitoring, collation and interpretation are lengthy and complex.
- The return of questionnaires is spasmodic
- Need non biased, non assisted questions
- Fear of putting their opinions in writing
- No direct guidance and assistance available
- Seen as an invasion of privacy or a threat
- Some "side information" may not be gathered - you only get answers to questions that you ask!!

Progress Reviews

The objective of a progress review is to:
- Check that the project is progressing according to the project plan
- Ensure that the Business Requirements and Objectives are being addressed
- Ensure that work is being undertaken to the agreed standards and quality
- Identify any deviations before they become harmful
- Re-schedule and re-plan as required
- Assure the users and project team that project progress is appropriate and happening

There are several types of reviews commonly used:

Internal Reviews:
- Client and Users - at the end of each stage or phase, and as appropriate or requested, the clients and users should be afforded a formal review and walk through of the project progress and objectives.
- Post Implementation - a formal review will identify any weaknesses, omissions and potential for system enhancements.

External Reviews:
- Quality - the Quality team will require to undertake a number of quality audits and reviews to assure themselves that the quality standards and plan is being adhered to. Quality Reviews will happen at the end of each stage, prior to a major deliverables, and after formal testing.
- Audit - the client's audit function will no doubt request a formal review of the system prior to implementation.

Project Reviews:
- Stage Review -it is helpful for the project team, led by the Project Manger, to review the project plan and progress against it. This ensures that all are aware of the plan and that it accurately reflects work to be undertaken.
- Project End Review - a formal review will not only highlight lessons to be learnt, but also reap valuable statistics and data for the next project

Project Inception and Set up

Project Set up

There are four main concerns to be addressed at Project Inception. These are:
- Why are you undertaking the project? - The Business Requirement and Business Case need to be ascertained, defined and agreed.
- What is the project team expected to do? - obtaining the Terms of Reference for the project team, how, when and where they are expected to work. What they are expected to achieve. What they are to be paid, and under what circumstances.
- What environment should the project work within? - Setting the Project Baselines. The costs, benefits, requirements, risk, budgets, time frame, resources etc. What is the scope of the project, what does it not involve?
- What is the project expected to achieve? - Objectives of the project

The above objectives will be consolidated and presented in the following reports:
- Terms of Reference
- Statement of Scope and Objectives
- Project Charter

The definition, production and agreement of these reports are by necessity an iterative one, the first two reports feeding into the Project Charter. Considerable consultancy will be needed, together with the build up of an effective Client Relationship. The production of the reports for agreement will probably undertake several versions, until they are finally agreed. This is to be expected, and is part of obtaining the formal definitions necessary to start the project.

Objective setting

A good set of objectives is:
- Relevant, appropriate and realistic
- Definable and controllable
- Clear, specific, attributable and measurable

The objectives of a project are difficult to define, but once done and disseminated to the clients, users and project team they are generally well understood. What is difficult is to understand how their individual tasks and achievements can contribute to these objectives.

The solution to this problem is to cascade the objectives down into the project team and users. The top-level objectives should be split into sub-objectives that will make up these top-level objectives. Lower level objectives should then be defined to make up these sub levels and then so on until each group and individual team member has their own objectives and tasks to achieve them.

Table 7 Objectives

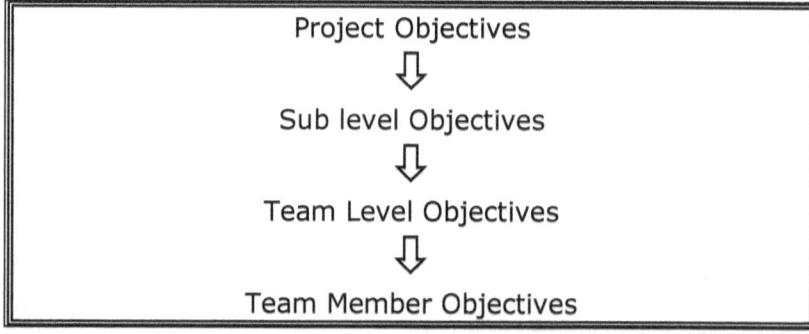

Project Charter

The Project Charter is an important document, which sets up the project, and the project teams abilities. Its formal acceptance marks the official start of the project and moves the project onto Project Inception and project planning. It has the following characteristics:
- Acts as a sales overview of the capabilities of the project team
- Ascertains and documents the aims and scope of the project
- Obtains formal agreement to Project Inception
- Schedules the project within the existing workload
- Formally starts the project
- Written as a short, snappy sales overview
- Produced by the Project Manager
- Based on discussions with management/client
- Presented to the client
- Reviewed with client management, where it may be subject to several reviews by different sanctioning boards, e.g. for budget, for technical approval etc.
- Rewritten and adjusted as necessary or requested
- Gain client Sign Off to proceed to Project Inception ☑

- The project is prioritised and scheduled as appropriate
- Used to inform project staff of Project Inception

Communication Reports

Terms of Reference

Guide lines for usage
Terms of Reference (TOR) - A formal document produced just before Project Inception, which defines the scope of the work to be undertaken by the consultants a project team, together with the project's time and resource allocations.

Terms of Reference (TOR) - Also used to define the description and boundaries of any work given to one person or a group of people, or a description of a task to be undertaken.

Recipients
- Steering Committee
- Project Sponsor
- Project Manager
- File copy

Report Contents
- Executive Summary: This section should provide in broad terms a description of the project, identifying the client and describing the client's current problems. The summary should indicate any proposed approach to be taken by the consultancy, together with an indication of the priority that has been assigned to the project.
- Description of work to be undertaken by consultants and project team. Describe in broad terms the scope, aims and boundaries of the work to be undertaken. Where work is to be undertaken by another group, their limits and interaction with the project team should also be defined.
- Time and budget allocation. Give estimates ± 50% at this stage
- Sign off by Project Sponsor. Agreement to the report contents and authority to proceed.

Statement of Scope and Objectives

Guide lines for usage:
A formal agreed document, produced just before project inception, which describes what the project encompasses, its background, what it hopes to achieve and any constraints on the solution.

Recipients:
- Steering Committee
- Project Sponsor
- Project Manager
- File copy

Report Contents

- Business Requirement: The formal definition of the problem faced by a business and the justification for requiring the system, an outline of the benefits which will accrue, the savings it will bring.
- Executive Summary: It should provide in broad terms a description of the project, its scope, objectives and constraints, identifying the client and describing the client's current problems.
- Project Scope: What the project encompasses and what factors or tasks are not to be addressed by the project. Include any phasing, with their time frames that may be appropriate. Include details of user involvement with the project.
- Project Objectives: The aims of the project and what it hopes to achieve. A list of definable milestones or deliverables would be the optimum constitutes of this section. Include details of how the objectives can be recognised, measured, agreed and signed off.
- Project Constraints: Include any constraints that the project must be implemented under - e.g. a particular methodology, technical platform, interfacing to another system, resource constraints. The constraints must be significant. Include suggestions as to how to overcome or negate these constraints, if possible at this stage. If not include overview of how the team intends to deal with them.
- Sign off by Project Sponsor: Agreement to the report contents and authority to proceed

Project Charter

Guide lines for usage:

The Project Manger with a two-fold aim produces this report:

- Written as a short, snappy sales overview of the project team's abilities
- A formal agreed definition of the scope and objectives and boundaries of the project.

FastTrack © Project Management

Recipients:
- Steering Committee
- Project Sponsor
- Project Manager
- File copy
- Overview to project team as appropriate

Aims:
- Act as a sales overview of the capabilities of the project team
- Ascertain and document the aims and scope of the project
- Obtain formal agreement to the project inception
- Schedule the project within the existing workload
- Formally start the project

Report Contents:
- Project Sponsor. Identify the client department and the prime contact points within the department or company, stating their involvement with the project.
- Business Requirement: The formal definition of the problem faced by a business and the justification for requiring the system, an outline of the benefits which will accrue, the savings it will bring.
- Executive Summary: This section should provide in broad terms a description of the project and its scope and objectives, identifying the client and describing the client's current problems. Where applicable, the summary should indicate any proposed approach to be taken by the consultancy, together with an indication of the priority, which has been assigned to the project.
- Definition of project. This section should provide in broad terms a description of the project, identifying the client and describing the client's current problems. An indication of the priority that has been assigned to the project should be indicated. Where applicable, the definition should indicate any proposed approach to be taken by the project team.
- Scope of Project. This section should document the limits of the project as currently perceived. In addition to identifying which aspects of the project will be referenced, the scope should also highlight any aspects that will not be considered.

- Project Objectives. The aims of the project and any identifiable deliverables. Where applicable project phases and delivery tranches should be identified and scheduled here. Any time scales given are to be quoted ± 50% accurate.
- Business Case. The justification for undertaking a project, defining the benefits that the system is expected to deliver, the savings it will accrue, measured against the cost of implementing and running the system and the risks that will be run.
- Feasibility. An assessment of the feasibility of the project, highlighting any known technical, operational, staffing or budgetary concerns. The obligations of the client, users and project team as they appertain to the feasibility of the project, should also be outlined.
- Identified Time Constraints. All known time constraints and requirements should be documented in this section. Any impact that any other project or sub-project will have on this project's time frame should be outlined.
- Ramifications of project. An overview of the possible ramifications of the project, especially with regard to the user environment, should be presented here. The obligations of the client, users and project team as they appertain to the project, should also be outlined.
- Cost Benefit Analysis. An approximation of the likely costs of the project as a whole should be provided, together with a list of any identified or perceived benefits that are to be gained from the project. Costings should be shown as either capital or revenue expenditure and where applicable revenue flows should be documented. Benefits should be tangible and intangible. All benefits and costs, qualitative and quantitative should be identified.
- Risk Analysis. An assessment of the perceived 'cost' of not implementing the project (the 'do-nothing" option), as well as implementation.
- Recommendations. Recommendations for the continuation or not of the project should be made.

- Steering Committee Authorisation. This section should include the signature of the Project Manager responsible for the project together with an appropriate place for the client to provide a formal acceptance of the Project Charter and its' contents. Where applicable a request for authorisation to proceed with the User Requirements Specification phase should be made.

Project Progress Report
Guide lines for usage:
Produced Project Manager on a weekly or monthly basis. It provides an up-to-date and informative report on the project progress highlights any possible problems and brings any decisions required to the senior user's attention.

Recipients:
- Steering Committee
- Project Sponsors
- Senior Users
- Quality Assurance Manager
- Project Manger
- File copy

Report Contents:
- Project Name
- Achievements during period. List any deliverables made, milestones passed, stages completed or other major accomplishments.
- Problems encountered or anticipated. Highlight any problems, how they were resolved, management decisions required to resolve them, or plans to solve them. Detail any impact on the project plan or resource allocation. Highlight anticipated problems
- Planned activity for coming period. Highlight at a macro level, work that is being undertaken and expected to be completed within the period between this report and the next.
- For Management Attention. Highlight any factors that you feel the senior managers need to be aware of. Indicate whether a decision or authorisation is required, its priority, impact and time frame.
- Project Manager's Authorisation

User Questionnaires

Guide lines for usage:
Questionnaires are a list of appropriate questions used within data gathering in the following cases:
- When a large statistical survey needs to be taken
- When job descriptions are required
- To obtain uniformity of approach
- To save time
- To obtain feedback from the users
- To make the users feel involved

They may also be used in face to face analysis - either as a prompt or as a control on the uniformity of questions asked.

Recipients:
- Relevant users
- Project Manager
- Developers
- Analysts
- File copy

Report Contents:
- Project Phase
- Activity, Data or Action
- Questions and either free flow or choice of answers
- Collator of information
- Time frame

Test Acceptance Report

Guide lines for usage:
An internal report produced by the programmer or testing team as a formal written report on the Unit or System Testing just completed.

Recipients:
- Project Manager
- Quality Manager
- Development
- File copy

Report Contents:
- List of Tests carried out
- Review of testing activities
- Test data used
- Results
- List of outstanding problems
- Recommendations
- Programmer's or testers sign off
- Quality Manager's authorisation
- Project Manager's authorisation

User Acceptance Test Report

Guide lines for usage:
An external report produced by the user testing team as a formal written report on the User Acceptance Testing just completed.

Recipients:
- Steering Committee
- Project Sponsors
- Senior Users
- Quality Assurance Manager
- Project Manger
- File copy

Report Contents:
- Project Name
- Deliverables
- Management Overview
- Review of testing activities
- Test data used
- Results
- Quality Manager's authorisation
- Project Manager authorisation

The Importance of Documentation

Documentation is the life blood of a project - it is used in planning, control, communication and support. It's always an unpopular task that gets in the way of the real "work". A balance between overcoming the common desire to leave system documentation until last, a common trait of programmers, and ensuring that the documentation does not strangle the project must be found.

Documentation must be undertaken as the programming, analysis and development is undertaken. Bit by bit is easier, more efficient and accurate.

There must be just enough formal administration to manage the project without burdening the project staff with too much administration.

Documentation is used as follows:

Project Planning:
- Personal - the team member's own planning, diary and notation documents.
- Project diary - meetings, phone calls, major contacts, commitments and important facts are recorded here. It also allows the Project Manager to be aware where his staff are and their work allocations at any one time.
- Public - the standard planning and planning control documents as outlined elsewhere.
-

Control:
- The control documentation shows the Project Manager what is happening to the development and implementation of the system, and aids its ongoing development.

Communication:
- Documentation allows information to be succinctly and efficiently broadcast to those that need it.

Support
- Documentation provides an important first line of user support.

- It allows efficient access to information on the structure and functionality of the system during periods of enhancement, upgrades, maintenance and auditing.

Analysis and Design - Standard Procedures

Whenever analysis and design is undertaken within a project there are a number of standard procedures that have to undertaken in order to ensure that a full picture is gained of the existing system, the proposed system and how to move from one system to the other. There follows a check list of factors to analyze.

Scope of the System:
- Potential Users - this should include main users, those who will need access to the information, the management stream and hidden users such as auditors:
  - Names.
  - Titles.
  - Outline Terms of Reference.
  - Type of access to system.
- Job structure - this should include a definition of the job, the skills needed, it's grade, it's status and it's dependencies:
  - Attractive jobs.
  - Unattractive jobs.
  - Badly designed.
  - Duplication of effort.
  - New jobs or tasks.
  - Obsolete jobs or tasks.
- Informal activities - the routines and procedures people employ to bypass a task, level of management or staff member or because a requirement is missing from the existing system
  - Description of activity.
  - What jobs are involved?
  - Reason for informal activity.
  - Impact.
  - Possible solution.

- System peculiarities - matters peculiar to the users or system
    - Jargon.
    - Abbreviations.
    - Unusual structure.
    - Industry or job specific customs or requirements.
    - Company standards.

Design of the System:
- Factors to be documented of the logical design (the user's outputs, inputs and processes):
    - Procedures: Flow Charts
    - Data: Data Flow Charts
    - Documents: Document Flow Diagrams
    - Entities: Entity Relationship Diagram,
    -                  Entity Life History

- Factors to be documented in the physical design - (the data files, program modules and supporting s/w, inc. operational requirements, security levels and authorized users):
    - System   System Specification
    - Program   Program Specification
    - Functionality   Functional Specification

- User interface and communication- this should include how they hold dialogues with the system, its help facilities and some feedback on the user's performance and actions.
    - Content
    - Format
    - Feedback

System Environment
- Workstation - its design will affect the confidence, happiness and efficiency of the user.
  - Format
  - Aims
  - Restrictions
- Work place and Hardware site - its design will affect the efficiency, performance and security of the system.
  - Format
  - Aims
  - Restrictions
- Data storage -it's design will affect it's accessibility, acceptability and security
  - Format
  - Restrictions

System Support and Maintenance

Full analysis and proposals must be made of the following to ensure efficiency, acceptability and security:
- Housekeeping
- Auditing
- Disaster Recovery
- Training
- Documentation
- Support and Secondary Support
- Maintenance
- Enhancements and Upgrades

Development and System Support

Documentation allows for efficient access to information on the structure and functionality of the system during periods of enhancement, upgrades, maintenance and auditing.

A system is a complex environment, which during both development and live running undergoes a number of changes, the documentation of the system and the changes are important for the following reasons:
- A change, enhancement or upgrade may have an impact on many different parts and modules of the system, it is important that the ramifications of any change is known.

- The documentation will define the programming methodology and format used.
- It is important to know what each part of the system does and how, in order that errors may be quickly found and addressed.
- Programmers move on to other positions - documentation provides a commonalty and continuation of support and information.

Without documentation a program and system soon becomes a mass of very expensive spaghetti that can not be cost effectively updated, enhanced, maintained or repaired.

User Support

User Support Documentation assists with the following:
- Training and technology transfer
- The users and user support function to become self supporting
- The motivation of the users to be able to use their system
- The users in a period of great change and uncertainty.

Technology Transfer

Technology transfer - the transfer of technological and system information from the project team to the users, managers and user support function is a very important part of a Project Manager's remit. No system can function on its own, and a project team must move onto other projects after implementation.

Technology transfer should ensure that the client company will, at the implementation of the system, have sufficient kills, abilities and experience to be able to manage and maintain the new system. The transfer should include users, line managers, senior managers, operational, maintenance and support personnel

A common cause of failure of a project after implementation is the inability or non readiness of users to fully utilize and support the new system.

It is also the professional obligation of every Project Manager to also transfer his knowledge to his own team members - otherwise how would the species be perpetuated?

Technological transfer should work in tandem and compliment training and usually takes the following formats:
- Shadowing - a user works alongside a project team member, undertaking the same tasks and responsibilities - shadowing the team member's work and learning "on-the-job". The transfer happens as an upward pull. Method demands a steep learning curve from the user and can be very daunting. Planning must take into account the lessening of efficiency of the team member. If the user can manage this work - he quickly becomes productive. Care must be taken to ensure that the user does not feel taken advantage of.
- Buddy System - a parallel pull. A project team member, befriends a user of the same work level, and teaches the person how to do his job. This involves a much gentler learning curve, but care must be taken to ensure that learning actually takes place. The lessening of the efficiency of the team member is less but the user does not become productive as quickly.
- Mentoring - an upward push. A senior project team member assists a senior user of a lesser seniority. The project member encourages, motivates and informally teaches the user, the requirements of the work. The lessons passed on are more those of experience, than that of "how to do a job". Good mentoring works both ways, with both parties gaining information and benefiting. The lessening of efficiency is not marked - as both parties have information to gain and supply.

Training

Care must be taken to ensure that training is efficient, timely, interesting and appropriate. Different types of users require different information and skills:
- End User
    - Input, output and navigation through the software
    - Production of standard reports
    - Basic housekeeping routines
    - Production of management reports
    - Office system skills
- Business User
    - Input, output and navigation through the software
    - Other end user skills as appropriate
    - Interpretation of the information supplied
    - Business skills as appropriate to the new system
    - Consolidation and collation of data
- Management
    - Input, output and navigation through the software
    - Office system skills
    - Management skills as appropriate to the new system
    - Interpretation of the information supplied
    - Management of the system
- Support
    - Input, output and navigation through the software
    - Support functions
    - Configuration Management
    - Housekeeping routines
    - Help Desk techniques
    - Disaster Recovery
- Maintenance
    - Input, output and navigation through the software
    - Maintenance functions
    - Configuration Management
    - Housekeeping routines
    - Help Desk techniques
    - Disaster Recovery

Project Documentation Reports

Analysis – it is so important to ensure that you document all of your analysis.

Analysis Reports

Document Flow Diagram
Guide lines for usage
Produced during Analysis stage to plot what happens to documents within the system. The correct use of this procedure will highlight:
- Double handling - documents handled more than once for the same task, or documents going to more than one entity.
- Loops in the system - a route taken more than once.
- Holes in the system - items or actions not covered or undertaken.
- Gaps in the system - lack of continuation of information flow.
- Inefficient handling - consolidation of one or more actions or data elements will speed the system.
- Lack of audit trail - no traceable path of actions taken.
- Poor security - lack of appropriate precautions.
- Poor housekeeping - inefficient housekeeping or storage routines.
- Inappropriate actions - illegal or unofficial actions, or those inappropriate to the task.
- Inappropriate actioner - task person has inappropriate skills or grade.

Recipients
- Project Manger
- Analysts
- Development
- File copy

Report Contents
A diagrammatic description of each document and how it moves (flows) through the existing and proposed system. Details of changes enacted upon each document. Produced for the systems analyst.

Data Flow Diagram

Guide lines for usage

Produced during Analysis stage to plot what happens to individual data elements within the system. The correct use of this procedure will highlight:
- Double handling - data handled more than once for the same task, or data going to more than one entity.
- Loops in the system - a route taken more than once.
- Holes in the system - items or actions not covered or undertaken.
- Gaps in the system - lack of continuation of information flow.
- Inefficient handling - consolidation of one or more actions or data elements will speed the system.
- Lack of audit trail - no traceable path of actions taken.
- Poor security - lack of appropriate precautions.
- Poor housekeeping - inefficient housekeeping or storage routines.
- Inappropriate actions - illegal or unofficial actions, or those inappropriate to the task.
- Inappropriate actioner - task person has inappropriate skills or grade.

Recipients
- Project Manger
- Analysts
- Development
- File copy

Report Contents

Similar to the Document Flow Diagram except that the flow of each piece of data appertaining to the documents **and** the system is described. Produced for the systems analyst.

Entity Relationship Diagram

Guide lines for usage

Produced during Analysis stage to plot what happens to individual entities (things) within the system. The correct use of this procedure will highlight:
- Double handling - data handled more than once for the same task, or data going to more than one entity, an entity being used twice for the same task.
- Loops in the system - a route taken more than once.
- Holes in the system - items or actions not covered or undertaken.
- Gaps in the system - lack of continuation of information flow.
- Inefficient handling - consolidation of one or more actions or data elements will speed the system.
- Lack of audit trail - no traceable path of actions taken.
- Poor security - lack of appropriate precautions.
- Poor housekeeping - inefficient housekeeping or storage routines.
- Inappropriate actions - illegal or unofficial actions, or those inappropriate to the task.
- Inappropriate actioner - task person has inappropriate skills or grade.

Recipients
    Project Manger
    Analysts
    Development
    File copy

Report Contents

A diagrammatic description of each small part (entity) of the system interacts with other parts of the system. Produced for the systems analyst.

Entity Life History

Guide lines for usage

Produced during Analysis stage to plot what happens to individual entities (things), over their life span, within the system. The correct use of this procedure will highlight:
- Double handling - data handled more than once for the same task, or data going to more than one entity.
- Loops in the system - a route taken more than once.
- Holes in the system - items or actions not covered or undertaken.
- Gaps in the system - lack of continuation of information flow.
- Inefficient handling - consolidation of one or more actions or data elements will speed the system.
- Lack of audit trail - no traceable path of actions taken.
- Poor security - lack of appropriate precautions.
- Poor housekeeping - inefficient housekeeping or storage routines.
- Inappropriate actions - illegal or unofficial actions, or those inappropriate to the task.
- Inappropriate actioner - task person has inappropriate skills or grade.

Recipients
- Project Manger
- Analysts
- Development
- File copy

Report Contents

A diagrammatic description of each small part (entity) of the system interacts and changes over the life span of the system. Produced for the systems analyst.

System Reports

Program Documentation

Guide lines for usage
- Included within the program as appropriate
- also separate report

Recipients
- Project Manager
- User Support Function
- File copy

Report Contents
- Overview
- Data structure
- Program structure
- Program "Calls" and "Called by"
- Files Used
- Screen standards
- Functional Description
- Benchmarks

Data Dictionary

Guide lines for usage
- To maintain consistency of variable and field usage and handling.
- To ensure that development methodology is followed with respect to variables and fields
- Used by the development, maintenance and support functions

Recipients
- Included within the program
- User Support Function
- File copy

Report Contents
- Details of all the variables and fields used within the program, modern RDBMS's produce the data dictionary automatically.

Management Reports

Management Overview
Guide lines for usage
- To give senior and line managers an overview of the benefits, functions and use of the new system
- To advise them of the reports and information available to them from the new system
- To enable them to fulfil their job function whilst making best use of the new system
- To enable them to advise upon and manage their staff's optimum usage of the new system

Recipients
- Project Manger
- Project Sponsor
- Senior Users
- File copy

Report Contents
- System overview
  - Overview of the objectives of the system
  - The role of the system within the business environment.
  - The benefits and demands of the system
  - The functionality of the system.
  - Information available
  - Output reports, screen shots etc.
- User Overview
  - Staff involvement with system
  - Day to day running and support
  - Contact points
- The Way Forward
  - Any future enhancements of developments

User Documentation

Help Screens

Guide lines for usage
To assist the user with the fast and efficient navigation around the system and its optimum use.

Recipients
- Included within the program
- User Support Function
- File copy

Report Contents
Screens of context sensitive guidance, "do-nexts" and help, that is accessed from a hot key - usually F1

User Handbook

Guide lines for usage
Gives full details for the day to day user running and usage of the system, together with immediate error identification and help contract points.

Recipients
- Project Manager
- Users
- User Support Function
- File copy

Report Contents
- System overview
  - Overview of the objectives of the system
  - The role of the system within the business environment.
  - The benefits and demands of the system
  - The functionality of the system.
- Information available
  - Output reports, screen shots etc.
- User Overview
- User involvement with system
  - Day to day running and support
  - Contact points
- Using the system

- Full details of the following:
  - Data input
  - Data handling
  - Data output
  - Keyboard layouts
  - Interfaces with other systems
  - Daily, weekly and monthly routines
  - Error Reports
  - Audit Reports
- Housekeeping
  - Daily routines
  - Daily maintenance
- Help and Support contact points
  - Full contact details for Help-Desk support

User Guide

Guide lines for usage
A more concise overview of the User Handbook, used for day to day reference and quick identification.

Recipients
- Project Manager
- Users
- User Support Function
- File copy

Report Contents
- Output reports, screen shots etc.
- Data input
- Data handling
- Data output
- Keyboard layouts
- Interfaces with other systems
- Daily, weekly and monthly routines
- Error messages
- Help and Support contact points

Procedures Guide

Guide lines for usage
Used to describe the procedures required to be in place to fully utilize the new system.

FastTrack © Project Management

Recipients
- Project Manager
- User Management
- Users
- User Support Function
- File copy

Report Contents
- System overview
- Procedure overview
- Flow chart of procedures
- Contact for Help

Maintenance and Support

Procedure Documentation

Guide lines for usage
Used to describe the procedures required to be in place to fully maintain and support the new system.

Recipients
- Project Manager
- User Management
- Users
- User Support Function
- File copy

Report Contents
- System overview
- Procedure overview
- Flow chart of procedures
- Contact for Help

System Documentation

Guide lines for usage
Used to describe the system to enable it to be fully maintained and supported.

Recipients
- Project Manager
- Users
- User Support Function
- File copy

Report Contents
- System overview
  - The role of the system within the operational environment.
  - The functionality of the system.
- System hierarchy
  - A top-down analysis of the system showing all procedures and functions used in the system
- Units in system
  - List all high level functions and procedures alphabetically detailing each unit's purpose.
- Files in system
  - List all files and databases alphabetically detailing file purpose.
- Appendices

User Support Documentation

Operations Guide

Guide lines for usage
Used to fully describe the operations functions to be performed in order to fully support the user.

Recipients
- Project Manager
- User Support Function
- File copy
-

Report Contents
- System overview
  - An introduction to the system which defines the role of the system within the operational environment.
- Installation
  - Guide to the installation of the system
- System hierarchy
  - A top-down analysis of the system showing all procedures and functions used in the system
- Operational guide by function
  - For each of the functional areas a full description of the module should be provided together with a description of the operation of the module. Where appropriate the descriptions should be supported with screen-shots.
- Occasional maintenance
  - Where applicable any occasional maintenance requirements (e.g. Yearly rollover or static data creation etc.)
- Configuration
  - Any configuration requirements
- Help Desk details

Installation and Maintenance Manual
Guide lines for usage
Full details of all installation and maintenance functions.

Recipients
- Project Manager
- User Support Function
- File copy

Report Contents
- System overview
  - An introduction to the system which defines the role of the system within the operational environment

- System hierarchy
  - A top-down analysis of the system showing all procedures and functions used in the system
- Installation
  - Full and comprehensive guide to installation of the system
- Maintenance
  - Full and comprehensive guide to the maintenance of the system
  - Where applicable any occasional maintenance requirement (e.g. Yearly rollover or static data creation etc.)
- Configuration
  - Any configuration requirements
- Help Desk details

Jargon Buster

Acceptance Test	Testing undertaken by the User that determines whether the system meets the requirements and specifications of the project
Acceptance Test Report	A document outlining the details of the Acceptance Test
Activity	An action within the project that has an aim, a start and finish and one or more resources allocated to it.
Analysis	Detailed and formal investigation - usually of actions and factors that will constitute a system
Analyst/Programmer	Project team member who cuts code and writes the software. He also undertakes system analysis work
Application System Manager	Senior member of project team, in charge of ensuring production, implementation and integrity of all software except the operating and comms software.
Baseline	A formally defined and non moveable or changeable starting point. It is used to measure variances in quality, risk, budget, resources etc. Also used when investigating problems within a project. The baseline is the known entity. The baseline is also used when releasing a sub-system or system, so that a know version can be released and all other changes and enhancements can be released as subsequent versions.
Benchmarks	Detailed numeric and qualitative actions and abilities that the system must possess, before it will be accepted by the users.
Benefits	Advantages both quantitative and qualitative that accrue from an action, system or option.

Budget	The amount of money allocated to a project. It is usually separated into amounts per resource or action per period.
Business Analyst	The member of the team that will analyze and document the existing and proposed system for the benefit of the project team and the user. He will ensure that all requirements, options, problems and advantages are fully known and formally documented.
Business Case	The justification for undertaking a project, defining the benefits that the system is expected to deliver, the savings it will accrue, measured against the cost of implementing and running the system and the risks that will be run.
Business Requirement	The formal definition of the problem faced by a business and the justification for requiring the system, an outline of the benefits which will accrue, the savings it will bring.
Capital	Capital costs are one-off costs on significant and tangible items, e.g. a computer.
Case Tools	A set of modules, supplied with some programs, or as a stand alone piece of software, that assist the analysts and programmers in formally designing and documenting the program. They are often part of modern methodologies.
Change Control	A formal method of documenting and controlling the changes to requirements requested during the life of a project.

Change Management	Change Management (as opposed to Change Control) is the management of the movement from one technical environment to another and/or from one system to another.
Client Relationship	The very important dialogue and correspondence with the project sponsor and senior users that ensures that the project runs smoothly.
Client-Server Architecture	A technical hardware and software configuration that consists of a mini computer (server) with PC.'s (clients) networked to it. The server undertakes the number crunching and database work and the client undertakes the more immediate data analysis. This set-up allows for the power of the mini computer and the flexibility of the P.C. to be fully utilized in a robust environment. PC.'s are also very familiar to many users.
Communications (comms)	The electronic connection, usually over telephone lines or glass fibre lines.
Configuration Management	The system used to identify, record, control, monitor and maintain master copies of all technical components created by the project.

Controls	The methodology explains how to undertake a project. Controls are used to ensure that the project is proceeding as required. They also form the basis of the reporting structure both internally and externally to the project team. Factors to be controlled within a project are: quality, risk, benefits, change, resources, budgets, project progress against the project plan, time, deliverables and software versions A baseline for each factor is established, from which variances and their ramifications are monitored and measured.
Data	The information within the system.
Data Dictionary	Details of all the variables and fields used within the program, modern RDBMS's produce the data dictionary automatically.
Data Flow Diagram (DFD)	Similar to the Document Flow Diagram except that the flow of each piece of data appertaining to the documents *and* the system is described. .Produced for the systems analyst.
Deliverables	A tangible, important part of the resulting system, the implementation of which marks a milestone within the project plan. Major Deliverables are often part of the contract between the project team and the users. All deliverables should feature within the project plan. An example is a major report, a software module or a computer sub-system.
Dependency	A constraint on the sequence and timing of a task within the project plan. It signifies that one task can not be completed or started before the completion or start of another.

Desk Top Publishing (DTP)	A software package that allows for graphics, newspapers, magazines, training material, slides etc. to be produced.
Document Flow Diagram	A diagrammatic description of each document and how it moves (flows) through the existing and proposed system. Details of changes enacted upon each document. Produced for the systems analyst.
Dos	Operating system for Pac's
Entity	A "thing" with a definite place, boundary and existence - e.g. store
Entity Life History (ELH)	A diagrammatic description of each small part (entity) of the system interacts and changes over the life span of the system. . Produced for the systems analyst.
Entity Relationship Diagram (ERD)	A diagrammatic description of each small part (entity) of the system interacts with other parts of the system. Produced for the systems analyst.
Error Control	An internal control mechanism and means of communication, used to capture, notify, resolve and document errors found during Quality Reviews, Unit, System and User Testing.
Error Report	Formal documentation outlining errors in programs and a method of formally signing off their resolution

FastTrack © Project Management

Executive Summary	Part of a long or complex document, found at the front, which gives n overview of the whole document, in sufficient detail to facilitate and executives understanding of the contents. It is used where the executive in question may not have the time or need to read the whole document. It should provide in broad terms a description of the project, identifying the client and describing the client's current problems. Where applicable, the summary should indicate any proposed approach to be taken by the consultancy, together with an indication of the priority which has been assigned to the project.
Flow Chart	Formal documentation, sometimes drawn informally, that outlines how information or documentation moves from one group or department to another. Used to outline a system.
Functionality	What the system does and how
Functional Specification	Formal documentation detailing the functionality (what it does) of a system.
Gantt Chart	A graphical way of project planning- shows actions over time-bars.
Hardware	The "hard" part of a computer system - the computer, peripherals, screens, Pac's etc.
Help Desk	A nominated area, often off-site, which acts as the first line support for problems and system errors. Often 80% of problems can be resolved by the Help Desk. The function escalates non- resolved problems to second, third or fourth level of support, depending on its severity.

Help Screen	Part of the software, such that if assistance or explanation is required by the user, a specific key (usually F1) can be hit to obtain context sensitive assistance.
Hot Line Support	Similar to a Help Desk, but usually refers to a telephone and /or modem connection. More immediate than a Help Desk and problems are usually resolved with the Helper talking to the User as he corrects the problem. Modem access is used when the Helper has to assist.
Housekeeping	The day to day routines that keep a computer system running- e.g. backing up, de-fragmentation, configuration management, archiving.
Implementation Team	The Implementation team, led by an Implementation Manager, are responsible for the installation of hardware, software and comms within the prescribed site.
Income	Income costs are often for intangible items and expected to be recurring, often beyond the life of the project, e.g. hardware maintenance.
Information Technology (IT)	Computing, the hardware, software, comms and profession.
Installation Manual	Detailed information on how to install the software. The manual is aimed at the skilled or informed user.
Interfaces	The method that one program module interacts with another. Modern RDBMS's have standardized interfaces.
Librarian	The role that controls the issue, receipt, storage and destruction of all reports, documents and software produced by the project.
Log Number	A unique number that identifies a document. Usually used in version, error and release control.

Logical Data Design	The formal documentation of the design of the proposed software. It models the data structure within the system.
Maintenance	The support and care of hardware, software and documentation to enable them to function effectively and efficiently. It also often includes the implementation of any upgrades.
Maintenance Manual	The documentation supplied to the maintenance team that outlines all the regular and ad hoc actions that are needed to maintain the system or product.
Management Overview	A brief synopsis, usually at the beginning of a report, that outlines the pertinent or most important part of the report, as it would appertain to a management decision or information stream.
Methodology	A structured and formal method of planning, managing, controlling and reporting upon a project. The methodology is used as a common language amongst the project team. It contains modules that define how to approach each stage of the project.
Milestones	Major events or deliverables within the project plan.
Module	A discrete section or part of the software, with its own functionality, that can operate on its own, but which also forms an integral part of a larger system.
Network	The formal method by which hardware and peripherals are connected together by communication lines, in such a way that hardware and software resource sharing can be undertaken.
Non Recurring Costs	One off costs not expected to occur again.

Open System	A system on a technological platform that is independent of a hardware or software vendor.
Operations Guide	A formal manual that outlines the various methods and actions needed to operate a hardware configuration,
Operations Manual	The same as the Operations Guide but with much greater detail.
Parallel Running	The period when the old system is run at the same time as the new system, using the same data. The purpose is to compare the output and performance of both systems. It is used in bench marking, testing and adherence to user specifications.
Pert Chart	A project planning technique, concentrating on time, which employs statistical analysis to estimate the probability of meeting target dates.
Phase	A division of the project plan, which denotes a group of complimentary project tasks - e.g. training
Physical Data Design	The conversion of the Logical System Specification into a design suitable for the proposed physical environment e.g. Paper design to a design specific to the proposed technical architecture.
Post Implementation Review	A formal review of the system, undertaken six to twelve months after implementation to check that the system is working smoothly and meeting user needs.
Procedure	The formal method that a group of actions must follow in order to achieve something.
Procedure Documentation	The documentation that details the methods, actions and procedures that must be followed to ensure that the system is effectively and optimally employed.
Procedures Guide	The manual given to the users that outlines

Program Specification	Formal documentation describing the details of the required software program
Programmer	The Programmer cuts code, writes the software and designs and writes the software modules and interfaces as required. He is also responsible for producing the Data Dictionary, Program Specification, System Documentation and Version Control Documentation.
Project	A Project is a group of defined and linked tasks that allow, using a formal methodology and a series of controls, a pre-determined business requirement to be fulfilled, by a series of formal deliverables within the constraints of time, resources, budgets and quality.
Project Charter	A formally agreed document, produced just before project inception that: Acts as a sales overview of the capabilities of the project team Ascertains and documents the aims and scope of the project Obtains formal agreement to the project inception
Project Closure	The formal ending of the project. It always includes the formal signing of the User Acceptance and often includes a review of actions taken and future options available.
Project Control Board	A group of users from the executive, senior management and senior technical echelons. It always includes the project sponsor or his nominated representative. The Board are empowered to make decisions as to the direction of the project, authorize changes and payments and sets the initial Terms of Reference and Statement of Scope and Objectives.

Project Control Office	The Project Control Office (PCO), working for, and very closely to the Project Management Team, has the following five functions: Communication Centre Reporting Controlling Secretarial and clerical Librarian
Project Control Officer	The senior manager in charge of the Project Control Office. He works closely with the Project Manager and is usually in charge of the project planning and control reporting.
Project Inception	The formal start of a project when all the requirements and project baselines are defined.
Project Initiation Document	Records the formal start to the project, it is prepared by the Project Manager and approved by the Control Board or Steering Committee. It is used to outline the project, identify any possible problems and raise issues that may need future resolution.
Project Library	The storage area for the hardware, software and documentation emanating from the project. The library should always be off-site and secure. The Project Control Office often has a set of documentation for immediate reference.
Project Manager	He leads, manages and controls the entire project team. He also designs, writes and controls the project plan and manages the client relationship. In smaller projects he undertakes the Technical Coordinator and Application Software Managers role, but never the Quality Manager's role.

Project Plan	The formal information and controlling documentation that details the resource allocation within a project to ensue that the objectives and requirements of the project are obtained.
Project Planning	The act of producing, reviewing, updating and controlling the project plan and the associated resources.
Project Review	An internal project control conducted on a regular basis, usually weekly. It is a formal report used to collate information of progress, problems and outstanding matters. External information, e.g. product information, can also be disseminated. The report is used to update the Project Plan and is used as formal documentation on project progress. By its nature it is not used externally to the project.
Project Sponsor	The person or persons who undertakes responsibility for the budget and cost of resourcing a project. He also undertakes the promotion of the project concepts and aims within the senior management of his company.
Prototype	A model of part or all of a system, presented to the client for his opinions and comments, in advance of full development.
Quality	An oft forgotten but very important part of the deliverables of a project. Quality measurement is one of the important controls within a project. A quality system is error free, well structured and defined, complete, comprehensive, reliable, easy to maintain and enhance, flexible, efficient and well documented. British Standard 4778 quotes "...the totality of features and characteristics of a product or service which bears on its ability to satisfy a given need."

Quality Assurance	The establishment of procedures and agreed standards for quality review and auditing.
Quality Control	The examination and control, to an agreed standard, of the products of a project.
Quality Manager	The senior manager in charge of producing and ensuring adherence to the Quality Plan.
Quality Plan	An agreed documented statement of the methods and procedures to be used within the life cycle of the project to ensure that a quality product and system is produced. It outlines the standards to be met by each product and deliverable in quantitative and qualitative terms.
Quality Review Team	The team, led by the Quality Manager, which undertakes the independent quality reviews. Not a member of the project team although they work closely together.
Rapid Application Development (RAD)	A method of development, often used in package applications, of producing an incremental series of prototypes for user analysis, comment and approval, which are then assimilated into the development cycle.
RDBMS	Relational Data Base Management System - a set of databases that allows the separate databases to interact with each other in a three dimensional way.
Recurring Costs	Those costs that are expected to recur within and after the project.
Release Control	A formal procedure of controlling the release and implementation of hardware configurations, software versions and documentation. It ensures up to date versions.
Resources	People, equipment, office space, computer time, expert time, etc. which are to be availed the project.

RFC	Request for Change to the specification of the system. Always used by programmers to formally document a change. Can be used by the analysts or users for smaller changes (large changes use the Change Control). It requires the Project Manager's signature.
Risk	The measurement of benefits or problems that may accrue from undertaking an action, design or configuration. The measurement may be quantitative, qualitative or opportunity.
Risk Analysis	The formal statistical or analytical investigation and measurement of Risk
Site Preparation (site prep)	The getting ready of an area for the imminent installation of hardware or implementation of software.
Software	Programs, and Operating Systems, sometimes the definition also includes consumables such as disc, tapes and documents.
Staffing Plan	A formal planning tool used to estimate project and user staffing requirements and usage. Costs and individual work allocation are often included.
Stage	Part of a project, denoting a group of complimentary project tasks, that has a formal, identifiable beginning and ending and is marked by a major deliverable - e.g. testing
Statement of Scope and Objectives	The definition of the objectives for a project, background, reasons and constraints on the solution.
Steering Committee	The User's senior management committee, responsible for the overall direction of the project, its Terms of Reference, Acceptance and implementation of the User's IT strategy.

Sub Project	A stage or tranche within the project may be so large or complex that it requires to be broken into sub-projects to ensure that all the tasks and activities can be appropriately recorded and actioned. If part of the project can be broken into a large discrete section that has no demanding dependencies upon the main project, and such that it can be more easily managed and controlled - it is often designated a sub-project.
Sub-System	A part of a major hardware or software system, that is capable of individual stand alone use, but which is an integral and integrated part of a larger system. e.g. a purchase ledger within a general ledger system
SWOT analysis	Investigation of the Strengths, Weaknesses, Opportunities and Threats, exposed by following an action
System	It may be hardware, software or both. It is the group of functions, actions and entities that make an entire entity, which exists to fulfil a user requirement.
System Documentation	The formal and agreed reports and documentation that describe a system, it's dependents, requirements and objectives
System Hierarchy	How a system is constructed, giving due consideration to the importance and/or timing of each individual module, sub system or entity.
System Specification	The formal agreed documentation that describes the system, its design, and constituents.
System Support	The group that undertake maintenance, upgrades and help facilities for an implemented system

FastTrack © Project Management

Systems Analyst	He interfaces between the Business Analyst and the Programmers. He will take the documentation produced by the business analysis and document it into a format suitable for the programmer to produce a system.
Task	An action, with a well defined entity, which must be undertaken in order to achieve or produce something. Tasks are often logically linked by action or time.
Technical Architecture	The hardware and software structure, how it in interlinked and interfaced. It also applies to the technology and design used to construct it - e.g. open systems
Technical Coordinator	The Technical Coordinate, works only on the hardware side of the project if there is an Application Software Manager in the team, otherwise he undertakes both software and hardware coordinating. He is responsible for making sure that the hardware and/or software components of the system are installed and implemented in the right place at the right time. This would encompass liaising with site preparation, testing, programmers, users and clients. This person needs to be organized, determined and tactful.
Terms of Reference (TOR)	A formal document produced just before Project Inception, which defines the scope of the work to be undertaken by the consultants a project team, together with the project's time and resource allocations. Also used to define the description and boundaries of any work given to one person or a group of people, or a description of a task to be undertaken.

Test Acceptance Report	The formal agreed documentation that accepts the results of a testing session.
Test Pack	A group of data that has a known, formally documented end result. It is used to give consistency in a series of tests and also to negate the necessity of testing within a live environment or on live data.
Test Plan	The formal agreed plan of what, when and how testing will take place.
Testers	The members of the project team and/or users that undertake the testing
Testing	Consists of the formal checking of an entity against the agreed expected results: There are three types.... *Unit testing* is conducted by the programmer responsible for production of the unit under test; generally the person who has programmed the unit. The purpose of *system testing* is to qualify the system as a whole, by testing the menu structures, user and program interfaces, screens, software modules, etc. The purpose of *user acceptance testing* is to qualify the system as a whole, to the satisfaction of the Project Sponsor, ensuring that the Project Goals have been met and the User Requirements satisfied.
Time	That of which there is never enough in any project.
Time Sheets	Formal agreed documentation that details time spent and costs of a team member. Can also be used for computer resource if appropriate
To-Do List	A living list of tasks, with a time element and priority level.

Trainer	Trainers, led by the Training Manager, design, write and undertake the training of the users, to ensure that they can use and maintain the new system after its implementation. Training usually starts as implementation begins.
Trainer Manager	He leads the training team
Tranche	A formal, time based element of a project plan. It always has a deliverable, and usually consists of several stages e.g. version 1.0 release is a separate tranche to version 2.0, hardware implementation in one area is different to another area.
User Acceptance	The formal agreement to accept an implemented system into the business environment. Responsibility for the system passes from the project team to the client.
User Acceptance Sign Off	The formal documented signature accepting responsibility for a system and signalling satisfaction with the project deliverables.
User Authorization	The formal authorization that either accepts a proposal or a change or allows the movement to the next stage or task within a project. It is an important control on User's Requirements. Often used as a check point of understanding.
User Documentation	A set of formal documentation written from the user's point of view that inform him of how to operate, maintain and upgrade an implemented system.
User Guide	An overview on a piece of hardware or software
User Handbook	A shorter précis of the User Guide
User Manual	Detailed information on a piece of hardware or software

User Questionnaire	Formal documented questions used within analysis to produce a standardized format. Used to enable statistical analysis of the answers and to ensure anonymity of answers.
User Requirement Specification (URS)	The formal, agreed documentation that sets out the full requirements that the system must meet.
User Support Documentation	Documentation that assists the users in the use, upgrade and maintenance of a system
Variance	The difference between an agreed baseline and the actual result.
Version	A release of a software program or hardware configuration that has its own unique functions and specification. Usually each succeeding version employs enhancements and improvements to the prior. Version X.y is the usual syntax used, where X signifies a release that includes enhancements or improvements. y signifies a minor version change to correct bugs and errors.
Version Control	Is used to ensure that a software program or hardware configuration is formally documented, recognized and controlled and that more than one version at a time, of a program is in existence within the live environment Is undertaken by the programmer or system development team leader prior to the installation of the software in the live environment.
Walk through	A "walk" by the product author of each individual step within the product. A "walk" by an analyst of the individual steps that a piece of data or document follows within an existing system

Work In Progress (WIP)	Work that has started but not yet been completed. Used in project planning to measure and estimate effort, resource usage and budgets
Work Plan	This is derived from the Project Plan and updated periodically. It is each individual member's list of tasks, responsibilities, quality and performance measurements and terms of reference.
Workflow Diagrams	Formal graphical documentation of the flow of work within a system.
Workshops	A controlled but free flowing way of obtaining analytical information from a group of people. Used to brain storm or think tank.

Index

Application System Manager, 21, 27, 166
Baseline, 12, 65, 166
Bid Manager, 2
Budget, 12, 43, 58, 60, 65, 72, 73, 74, 75, 77, 79, 90, 131, 167
budgets, 12, 36, 67, 70, 71, 74, 81, 94, 134, 137, 169, 175, 185
Budgets, 67, 69, 70
Business Analysts, 23
Business Case, 12, 38, 40, 118, 124, 137, 143, 167
Business Requirement, 11, 16, 38, 39, 40, 41, 42, 45, 56, 75, 118, 119, 124, 125, 131, 137, 141, 142, 167
change, 12, 13, 17, 19, 37, 42, 44, 53, 80, 81, 82, 83, 85, 86, 87, 88, 99, 100, 101, 102, 103, 111, 112, 115, 134, 150, 151, 169, 179, 183, 184
Change Management, 101, 130, 131, 133, 168
charge out, 31, 32, 71, 72, 74
Communicate, 20, 30, 58, 59
Configuration team, 27
Consultancy, 71, 73, 93, 101, 130
Control, 17, 20, 22, 25, 48, 49, 51, 52, 53, 58, 59, 61, 62, 75, 76, 77, 78, 79, 80, 82, 85, 86, 88, 89, 90, 92, 93, 94, 101, 103, 105, 106, 109, 110, 111, 112, 113, 117, 119, 123, 126, 147, 167, 168, 170, 175, 176, 178, 179, 184
Controls, 11, 13, 81, 82, 83, 95, 99, 169
Critical Path, 65
Data Conversion, 36, 53, 58
Data Flow Diagrams, 24, 47
deliverables, 11, 12, 35, 36, 41, 65, 69, 71, 75, 78, 81, 83, 102, 115, 134, 135, 141, 143, 144, 169, 173, 175, 177, 183
Deliverables, 12, 42, 44, 49, 52, 65, 69, 75, 104, 106, 113, 114, 146, 169
Document, 20, 24, 40, 42, 46, 47, 48, 50, 51, 58, 59, 60, 82, 85, 107, 108, 111, 117, 123, 149, 154, 155, 169, 170, 176
Entity Life History, 24, 47, 58, 149, 157, 170
Entity Relationship Diagrams, 24, 47
Executive Summary, 38, 40, 46, 60, 116, 120, 122, 126, 129, 140, 141, 142, 171
Feasibility, 17, 36, 38, 39, 40, 54, 58, 60, 63, 118, 125, 143

Functional Specification, 24, 46, 58, 61, 122, 128, 149, 171
Implementation, 19, 27, 36, 52, 54, 58, 73, 81, 93, 94, 99, 100, 101, 102, 111, 122, 128, 135, 172, 174
Implementation Team, 27, 172
Jargon Buster, 166
Lee Lister, 2
Legal Notice, 10
Maintenance Team, 28
Methodology, 11, 90, 173
Milestones, 42, 47, 65, 69, 75, 173
Plan, 20, 26, 27, 42, 43, 44, 47, 48, 50, 51, 53, 58, 60, 62, 64, 67, 68, 70, 75, 76, 78, 79, 80, 85, 86, 90, 100, 106, 108, 109, 111, 117, 119, 122, 123, 126, 128, 132, 133, 177, 178, 179, 182, 185
Programmers, 24, 25, 151, 181
Progress Reviews, 135
project, 2, 4, 11, 12, 13, 14, 15, 16, 17, 18, 19, 20, 21, 22, 23, 24, 25, 26, 27, 28, 29, 30, 31, 32, 33, 34, 35, 36, 37, 38, 39, 40, 41, 42, 44, 45, 50, 51, 53, 54, 56, 57, 58, 60, 64, 65, 66, 67, 68, 69, 70, 71, 72, 74, 75, 76, 78, 81, 82, 83, 84, 85, 86, 87, 88, 89, 90, 91, 93, 94, 99, 100, 101, 102, 105, 106, 108, 109, 110, 111, 115, 116, 117, 118, 119, 120, 123, 124, 125, 126, 130, 131, 132, 133, 134, 135, 136, 137, 138, 139, 140, 141, 142, 143, 144, 147, 148, 151, 152, 166, 167, 168, 169, 171, 172, 173, 174, 175, 176, 177, 178, 179, 180, 181, 182, 183, 185
Project, 3
Project Charter, 38, 58, 60, 117, 123, 137, 139, 141, 144, 175
Project Initiation, 16, 40, 58, 60, 117, 123, 176
Project Leader, 22
Project Manager, 13, 14, 16, 18, 19, 21, 22, 23, 24, 32, 43, 44, 50, 51, 57, 68, 71, 73, 75, 76, 77, 78, 79, 80, 82, 83, 86, 88, 100, 101, 105, 106, 107, 108, 110, 111, 112, 113, 115, 116, 117, 120, 121, 122, 123, 124, 126, 127, 128, 129, 130, 131, 133, 139, 140, 142, 144, 145, 146, 147, 151, 152, 158, 160, 161, 162, 163, 164, 176, 179
Project Meetings, 132
Project Methodologies, 35
Project Objectives, 27, 38, 51, 138, 141, 143
Project planning, 14
Project Planning, 36, 41, 54, 58, 64, 67, 81, 94, 147, 177

Project Reports, 37, 131, 132, 133
Project Scope, 38, 40, 118, 124, 141
Project Stages, 35, 36, 42, 69, 76
Project Strategy, 15
project team, 11, 12, 14, 18, 19, 21, 22, 25, 32, 33, 35, 37, 38, 39, 41, 42, 44, 60, 65, 69, 70, 71, 72, 75, 76, 82, 83, 84, 87, 89, 91, 93, 94, 102, 105, 109, 115, 116, 117, 118, 123, 124, 131, 132, 134, 135, 136, 137, 138, 139, 140, 141, 142, 143, 151, 152, 166, 167, 169, 173, 175, 176, 178, 181, 182, 183
Quality, 12, 18, 21, 23, 27, 36, 44, 51, 52, 53, 57, 58, 60, 62, 63, 75, 76, 77, 78, 79, 80, 84, 85, 87, 88, 89, 90, 91, 101, 106, 110, 111, 112, 113, 116, 130, 131, 135, 144, 145, 146, 170, 176, 177, 178
Quality Assurance, 87, 91
Release Numbering, 103
resources, 11, 12, 15, 17, 35, 36, 37, 38, 39, 41, 42, 54, 65, 69, 70, 71, 74, 77, 78, 81, 84, 85, 90, 91, 92, 93, 94, 110, 119, 125, 130, 131, 137, 166, 169, 175, 177
Resources, 11, 69, 178

risk, 10, 12, 15, 17, 39, 41, 65, 81, 83, 84, 85, 93, 137, 166, 169
Risk Assessment, 93
Risk Levels, 15
Security, 46, 95, 120, 122, 126, 128
Site Preparation, 28, 73, 179
Specialist Consultants, 23
Staffing Plan, 75
Statement of Scope and Objectives, 38, 58, 60, 117, 123, 137, 140, 175, 179
Steering Committee, 21, 39, 57, 85, 89, 117, 120, 121, 124, 126, 127, 130, 140, 142, 144, 146, 176, 179
System Specification, 24, 47, 48, 58, 121, 127, 149, 174, 180
System Support, 36, 54, 59, 150, 180
System Testing, 26, 50, 108, 145
Systems Analysts, 23
Systems Designer, 25
Tasks, 11, 65, 181
Technical Co-ordinator, 21
Technology Transfer, 46, 121, 127, 151
Terms of Reference, 33, 37, 38, 39, 40, 43, 58, 60, 72, 76, 117, 118, 123, 124, 137, 139, 148, 175, 179, 181
Testing, 27, 36, 44, 50, 51, 52, 53, 54, 56, 58, 63, 73, 78, 79, 80, 87, 88, 89, 106, 109, 111, 112, 116, 146, 166, 170, 182

The Bid Manager"., 2
Trainers, 24, 26, 183
Transition Planning, 70
Unit Testing, 26, 50, 107
User Acceptance, 27, 36, 51, 52, 53, 54, 56, 57, 59, 63, 79, 84, 85, 88, 91, 100, 105, 106, 109, 110, 111, 112, 114, 115, 116, 146, 175, 183

User Acceptance Testing,, 27, 51, 88
User Requirement Specification, 24, 37, 45, 46, 56, 58, 106, 115, 184
User Support, 36, 54, 55, 59, 62, 113, 151, 158, 160, 161, 162, 163, 164, 184
Work Allocation, 53, 68, 69, 72, 74, 76, 112

www.ingramcontent.com/pod-product-compliance
Lightning Source LLC
Chambersburg PA
CBHW020654220526
45464CB00001B/430